Amateur Radio Emergency Communications Guidebook

Dr. John A. Allocca, WB2LUA

Amateur Radio Emergency Communications Guidebook

Dr. John A. Allocca, WB2LUA
Northport, NY 11768
(631) 757-3919
john@allocca.com
www.allocca.com
www.WB2LUA.com

Copyright 2016, Updated 4/11/21

This guidebook may be reproduced, provided it is reproduced in its entirety without any changes

ISBN-13: 978-1530388400

Table of Contents

Amateur Radio Emergency Communications in the Community 5

Amateur Radio Emergency Service (ARES) 6

Radio Amateur Civil Emergency Service (RACES) 8

Principles of Disaster Communication 10

Working with Public Safety Officials and Agencies 12

National Traffic System (NTS) 13

Incident Command System (ICS) 14

Message Handling 16

Hurricane Intensity Scale 17

Estimating the Manpower Necessary to Service an Emergency Event 18

The Emergency 19

Radiogram 20

ITU Phonetic Alphabet 21

International Q Signals 22

U.S. Amateur Bands 24

Amateur Radio Call Sign Numerical Prefixes 26

Signal Reporting 28

Communications Net Procedure 29

CTCSS (PL) Tone Frequencies 31

Packet Radio 32

APRS (Automatic Packet Reporting Service) 45

Phase Shift Keying (PSK) Radio 51

Winlink 2000 55

Satellite Radio 62

Antennas and Propagation 70

H.F. Antenna Analysis 79

Emergency Power 87

Anderson PowerpolesR 93

D-Star and Programming 95

Narrow-Band Emergency Message System (EBEMS) - MT-63 Operating Instructions 123

Echolink and IRLP 128

Portable Antenna Systems 132

Amateur Radio Emergency Communications in the Community

Every major disaster throughout the entire world represents sudden local emergency conditions where loss of life, limb, property, necessary resources and even the ability to call for help have been forced upon people somewhere. When the news story breaks and we hear about it in the midst of our daily lives, the story is about the event itself and the extensive upset to life at the scene. However, somewhere in those initial reports, you usually hear that it was some local ham radio operator who was first able to re-establish communications and get out the call for help. They're usually first, they're usually there, and they usually get it done!

In our country, these reliable, highly trained, and dedicated amateur radio or "ham" radio operators are the same people you know as friends and neighbors. Amateurs they are, as they receive no pay or compensation for the services they eagerly provide in such times of crisis. The pure satisfaction of provisioning extremely effective civil emergency communications is their fulfilling reward. You'll recall that it was ham radio operators who provided the first communications downtown on 9/11 when the WTC disaster eliminated electric power, radio, television, and even NYC emergency communications were disrupted. Hams established communications within a few hours, while it was days before anything else approached normal. And that was right here at home!

Amateur Radio ("ham") Operators must be trained and skilled in many aspects of communications and radio technology in order to pass strict federal licensing examinations to earn their Federal Communications Commission issued licenses and radio "call sign." In very real terms, they are anything but amateur in the performance and utilization of their skills. They own and maintain their own radio equipment and are responsible for all aspects of the operation of their radio stations, whether it is from a fixed base location, a mobile station, portable station, or from aircraft or marine locations. Hams have built, orbited, and operated their own satellites since 1961, only 4 years after the world's first satellite, Sputnik, blazed the skies. Hams are for real, and they are an incredibly valuable asset to the world, all the time!

Why use Amateur Radio? The answer is simple and obvious, and it's because amateur radio equipment is independent of commercial radio services like telephones, cell phones, and even Police, Fire, and EMS service radio services, which are very limited in frequency and interoperability. Ham radio (Amateur Radio) is inherently frequency agile and readily portable, thus it is ideal for emergency dependability. Many hams are able to pick up and go, and set up communications on a moment's notice from almost anywhere. Many do just that for the enjoyment of it. You'll see hams in the parks and around towns providing supporting communications for public events like parades, marathon runs, etc. Such events are easy practice for hams, yet major events like the Boston Marathon and the New York Marathon critically depend on them because hams get the job done.

Amateur Radio Emergency Service (ARES)

The Amateur Radio Emergency Service (ARES) is a private volunteer organization of licensed amateur radio operators. It is not a part of any government organization. The only qualifications required are a valid FCC amateur radio license. ARES may assist private organizations, such as The American Red Cross, The Salvation Army, etc. ARES may also assist with community events such as marathon races. Only certified RACES personnel can assist government organizations, such as state, county, town, village, police, fire, EMS, etc. ARES is organized as follows:

National

- Advising all ARES officials.
- Setting and carrying out the League's policies.

Section

- Section Manager appoints the Section Emergency Coordinator (SEC).
- Section Emergency Coordinator (SEC).
- The Section manager is elected by the ARRL members in the section.
- The Section manager delegates to the SEC the section emergency plan.
- The Emergency Coordinator has the authority to appoint District and local EC's.

Local

- The local Emergency Coordinator (EC) is the key contact.
- Direct contact with the ARES volunteers and with officials of the agencies to be served.
- The EC is appointed by the SEC, usually on the recommendation of the DEC.

District

- In large sections, SECs have the option of grouping their EC jurisdictions into "districts"
- SEC appoints a District EC to coordinate the activities of the local ECs in the district.

Assistant EC's

- Assistant Emergency Coordinators (AEC) head up special interest groups or projects.
- AEC's are designated by the EC to supervise activities of groups or projects.
- AEC's provide relief for the EC.

ARES Operation During Emergencies and Disasters
- Operation in an emergency net requires preparation and training.
- Handling of written messages (traffic handling).

The ARRL Simulated Emergency Test (SET)
- Nationwide exercise in emergency communications, administered by ARRL Emergency
- Coordinators and Net Managers.
- ARES and the National Traffic System (NTS) are involved.
- SET provides the opportunity to discover the emergency communications capabilities.
- SET weekend is held in October, and is announced in QST.
- To find out the strengths and weaknesses of ARES and NTS.
- To provide a public demonstration to served agencies such as Red Cross and through the news media to the public.
- To help radio amateurs gain experience in emergency communications.

During the SET
- The "emergency" situation is announced and the emergency net is activated.
- Stations are dispatched to their positions.
- Designated stations originate messages to test the system.
- Test messages may be sent simulating requests for supplies.
- Tactical communications for served agencies is emphasized.

After the SET
- Critique session to discuss the test results and review good points and weaknesses.

ARES Mutual Assistance Team (ARESMAT)
- ARES members in an affected area may not be able to respond to ARES operation because of their own personal situations.
- Communications support must come from ARES volunteers outside the affected areas.

Radio Amateur Civil Emergency Service (RACES)

RACES is authorized by local, county, state, and federal emergency management agencies, under the direct control of the Federal Emergency Management Agency (FEMA) of the United States government. Amateur Radio Service provides radio communications during periods of local, regional or national civil emergencies.

As defined in the FCC rules, RACES is a radio communication service, conducted by volunteer licensed amateurs, designed to provide emergency communications to local or state civil-preparedness agencies. RACES operation is authorized by emergency management officials only.

To become a member of RACES, a licensed amateur radio operator must be officially enrolled in the local civil-preparedness agency having jurisdiction. Operator privileges in RACES depend upon the class of license held. In the event that the President invokes his War Emergency Powers, amateurs involved with RACES might be limited to certain specific frequencies (while all other amateur operation could be silenced). Originally, RACES was designed for wartime. It has evolved over the years to include all types of emergencies to government organizations, such as town, county, state, police, fire, EMS, etc. Only certified RACES personnel may assist government organizations and workers through the incident command system.

Dedicated RACES Operating Frequencies

1800-1825 kHz

1975-2000 kHz

3.50-3.55 MHz

3.93-3.98 MHz

3.984-4.000 MHz

7.079-7.125 MHz

7.245-7.255 MHz

10.10-10.15 MHz

14.047-14.053 MHz

14.22-14.23 MHz

14.331-14.350 MHz

21.047-21.053 MHz

21.228-21.267 MHz

28.55-28.75 MHz

29.237-29.273 MHz
29.45-29.65 MHz
50.35-50.75 MHz
52-54 MHz
144.50-145.71 MHz
146-148 MHz
222-225 MHz
420-450 MHz
1240-1300 MHz
2390-2450 MHz

Principles of Disaster Communication

Principles of Disaster Communications
- Keep the non-critical communications level down.
- If you're not sure you should transmit, don't.
- Study the situation by listening.
- Don't transmit unless you are sure you can help by doing so.
- Don't ever break into a disaster net just to inform the control station you are there if needed.
- Monitor established disaster frequencies.
- On CW, SOS is universally recognized.
- On voice, "MAYDAY" or "EMERGENCY" is universally recognized.
- Avoid spreading rumors.
- Authenticate all messages.
- Strive for efficiency.
- Select the mode and band to suit the need.

CW Mode
- Less non-critical communications in most amateur bands.
- Some secrecy of communications - less likely to be intercepted by the general public.
- Simpler transmitting equipment.
- Greater accuracy in record communications.
- Longer range for a given amount of power.

Voice Mode
- More practical for portable and mobile work.
- More widespread availability of operators.
- Faster communication for tactical or "command" purposes.
- Official-to-official and phone-patch capability.

Digital Modes
- Less non-critical communications in most amateur bands.
- Secrecy of communications - less likely to be intercepted by the general public with a scanner.
- Greater speed.
- Potential for message store-and-forward capability from within the disaster site to the "outside world."

● Provides the capability of "digipeating" messages from point A to point Z via numerous automatically-controlled middle points.

Working with Public Safety Officials and Agencies

Volunteers must be accepted by public-officials. Once accepted, they can to contribute in times of disaster. Acceptance is based on establishing a track record of competent performance in important activities. This may include parades, runs, and various other local events.

Police and fire officials tend to be very cautious and skeptical concerning those who are not members of the public-safety professions. This posture is based primarily on experiences where volunteers have complicated, and jeopardized, efforts in emergencies.

Volunteers need to demonstrate the reliability and clarity of amateur gear. Police and fire officials are very impressed to witness a roll call on a 2-meter repeater using a hand-held radio in the police or fire chief's office and having amateurs respond with full-quieting signals from locations where municipal radios are often ineffective.

As funding becomes less available, agencies are looking for volunteers. Relationships with served agencies are vitally important and valuable to radio amateurs. We provide them with communications. They provide us with the opportunity to contribute to the relief of suffering.

A detailed local operational plan should be developed with local agency managers. The plan should include the technical issues involving message format, security of message transmission, disaster welfare inquiry policies, etc.

National Traffic System (NTS)

The National Traffic System is a plan for handling amateur radio traffic. It is designed for rapid movement of traffic from origin to destination. It is also designed to train amateur operators to handle written traffic and participate in directed nets. The NTS consists of operators who participate for one or two periods a week, and some who are active daily.

Each net performs its function and only its function in the overall organization. To be an individual station in NTS, one must be issued certificates, and be appointed to the field organization's traffic handling position, entitled Official Relay Station.

Voice, CW, RTTY, AMTOR, packet or other digital mode is set up by the Net Manager or Managers concerned and the dictates of logic. There is only one National Traffic System, not separate systems for each mode.

Local nets are cover small areas such as a community, city, county or metropolitan area. They usually operate by 2-meter FM and use repeaters. A local net, or "node," may also be conducted on a local packet BBS.

Region nets cover a wider area, such as a call area. Regional nets consist of:

- A net control station, designated by the region net manager.
- Representatives from each of the various sections in the region, designated by their section net managers.
- One or more stations designated by the region net manager to handle traffic going to points outside the region.
- One or more stations bringing traffic down from higher NTS echelons.
- Any other station with traffic.

Area Nets are at the top level of NTS nets. Area nets consist of:

- A net control station, designated by the area net manager.
- One or more representatives from each region net in the area, designated by the region net managers.
- Stations designated to handle traffic going to other areas.
- Stations designated to bring traffic from other areas.
- Any station with traffic.

Digital Stations handle traffic among sections, regions and areas. These stations handle traffic by digital modes. They supplement the existing system, providing options, and flexibility in getting traffic moved expeditiously across the country, especially in overload conditions.

Incident Command System (ICS)

Almost all emergency government agencies have adopted the incident command system. It is a management tool that provides a coordinated system of command structure. Amateur radio operators should familiarized themselves with the system and how they may interface with government agencies that use the ICS.

The basic concept of the ICS is having a unified command. There is one person in charge of the emergency, the incident commander, who is totally responsible for everything that occurs in that emergency operation.

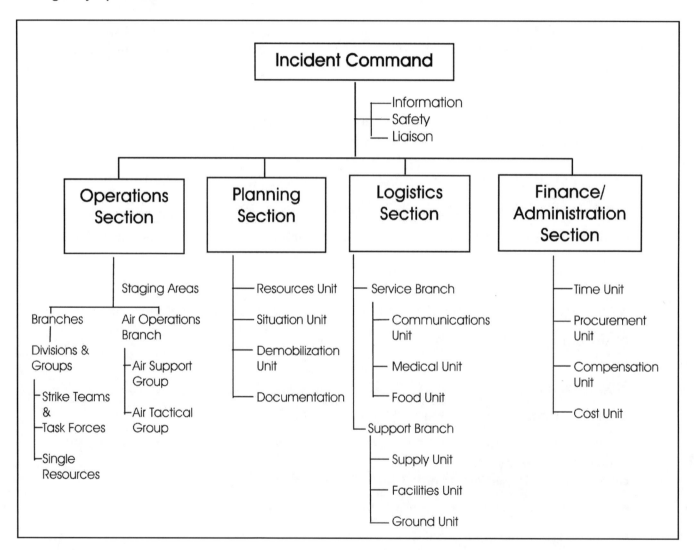

Command – Set objectives and priorities. Has overall responsibility at the incident or event.

Operations – Conducts tactical operations to carry out the plan. Develops the tactical objectives, organization, and directs all resources.

Planning – Develops the action plan to accomplish the objectives. Collects and evaluates information. Maintains the resource status.

Logistics – Provides support to meet incident needs. Provides resources and all other services needed to support the service.

Finance / Administration – Monitors costs related to incident. Provides accounting, procurement, time recording and cost analysis.

Incident Facilities:

1. Incident Command Post (ICP) – The location from which the Incident Commander oversees all incident operations

2. Staging Areas – Locations at which the resources are kept while awaiting incident assignment. Large incidents may have several staging areas.

3. Base – The location at the incident at which primary service and support activities are performed.

4. Camps – Incident locations where resources may be kept to support incident operations. There resources may not be immediately available.

5. Helibase – A location in and around an incident area at which helicopters may be parked, maintained, fueled, and equipped for incident operations.

6. Helispots – Helispots are temporary locations where helicopters can land and load and off load personnel, equipment, and supplies. Large incidents may have several helispots.

Message Handling

1. Speak in plain language.

2. Speak slowly and clearly.

3. Remain calm at all times.

4. If you have an emergency message, state the word "emergency" followed by your call sign.

5. If you have a priority message, state the word "priority" followed by your call sign.

   Emergency – any message having life and death urgency to any person or group of persons.

   Priority – any important message that has a specific time limit.

   Welfare – can be either an inquiry as to the health and welfare of an individual in the disaster area, or an advisory from the disaster area that indicates all is well.

   Routine – Most traffic will be routine in nature. In a disaster situation, routine messages should be handled last.

Hurricane Intensity Scale

Saffir-Simpson Hurricane Scale.

Category	Wind Speed	Barometric Pressure	Storm Surge	Damage Potential
1 (weak)	75 - 95 mph	28.94" +	4' – 5'	Minimal damage to vegetation
2 (moderate)	96 – 110 mph	28.50" – 28.93"	6' – 8'	Moderate damage to houses
3 (strong)	111 – 130 mph	27.91" – 28.49"	9' – 12'	Extensive damage to small buildings
4 (very strong)	131 – 155 mph	27.17" – 27.90"	13' – 18'	Extreme structural damage
5 (devastating)	155 mph +	less than 27.17"	18' +	Catastrophic building failures possible

The National Hurricane Center monitors 14.325 MHz and takes reports from Amateur Radio Operators during the storm. Hurricane season is June 1 to November 30.

NOAA Weather Radio:
162.400 mhz
162.425 mhz
162.450 mhz
162.475 mhz
162.500 mhz
162.525 mhz
162.550 mhz

Estimating the Manpower Necessary to Service an Emergency Event

One Person Per Shift

Assuming the following:
12 hour shifts
1 people per shift
A volunteer may only volunteer one out of every three days.

Therefore:
2 people are needed to cover 1 day
4 people are needed to cover 2 days
6 people are needed to cover 3 days

Conclusion:
6 people are needed to cover one assignment location.
12 people are needed to cover two assignment locations, and so on.

Two People Per Shift

Assuming the following:
12 hour shifts
2 people per shift
A volunteer may only volunteer one out of every three days.

Therefore:
4 people are needed to cover 1 day
8 people are needed to cover 2 days
12 people are needed to cover 3 days

Conclusion:
12 people are needed to cover one assignment location.
24 people are needed to cover two assignment locations, and so on.

The Emergency

Sometimes volunteers are called upon because emergency communications is needed immediately. Other times, volunteers are called upon to serve as back-up support in anticipation of losing total communications. Don't be discouraged if your services don't appear to have been useful. Having you there in place and ready to operate provides a very valuable service.

First Step

Before volunteering for emergency communications, be sure of the following:
- Family are safe and secure.
- Family has enough provisions, etc.
- Property is safe
- Monitor the designated frequencies, radio, and t.v.
- Contact your Emergency Coordinator or designee for instructions.
- Check batteries.
- Check medications if applicable.

Second Step

- Know and understand the volunteer handout.
- Do not take action until you are told to act.
- Be prepared to operate.
- Check all equipment and connections.
- Have pencil, paper, and radiograms ready.
- Obtain tactical frequencies.
- Check-in with your designated net or operations.
- Obtain tactical call sign if appropriate.
- Monitor all frequencies assigned to you.
- Notify net control operator if you have to leave.

Field Operations

- BE SURE TO ABSOLUTELY FOLLOW THE CHAIN OF COMMAND!
- If you are operating in the field, always keep a safe distance from any hazards.
- Keep yourself well hydrated (drink plenty of water).
- Take breaks and get rest when you can.
- Do not overexert yourself. Be aware of your own limitations.
- Do not overreact, become hysterical, or try to provide more help than is needed.
- DO NOT BECOME A VICTIM YOURSELF

Radiogram

Number _____

Precedence _____

Station of Origin _____

Place of Origin _____

Date Filed _____

Time Filed _____

To _____

_____

Telephone Number ( ) _____

Received at:

    Station _____

    Name _____

    Street Address _____

    City, State, Zip _____

Text _____

_____

_____

_____

_____

_____

_____

    From _____

    Date _____

    Time _____

    Sent to _____

    Date _____

    Time _____

ITU Phonetic Alphabet

A - Alpha
B - Bravo
C - Charlie
D - Delta
E - Echo
F - Foxtrot
G - Golf
H - Hotel
I - India
J - Juliet
K - Kilo (pronounced keelo)
L - Lima (pronounced leema)
M - Mike
N - November
O - Oscar
P - Papa
Q - Quebec (pronounced kaybek)
R - Romeo
S - Sierra
T - Tango
U - Uniform
V - Victor
W - Whiskey
X - X-ray
Y - Yankee
Z - Zulu

International Q Signals

QRA - What is your call sign?

QRG - Will you tell me my exact frequency (or the frequency of...)?

QRH - Does my frequency vary?

QRI - What is the tonal quality of my transmission?

QRJ - Are you receiving my transmissions poorly?

QRK - What is the intelligibility of my signals?

QRL - Are you/is the frequency busy? More!

QRM - Is there man-made interference to my transmissions? More!

QRN - Are you troubled by static or some other natural source of noise? (ok, cut the jokes :-) More!

QRO - Shall I increase power? More!

QRP - Shall I decrease power? More!

QRQ - Shall I send faster? More!

QRS - Shall I send more slowly?

QRT - Shall I stop sending? More!

QRU - Do you have anything for me?

QRV - Are you ready? More!

QRX - When will you call me again?

QRY - What is my turn?

QRZ - Who is calling me?

QSA - What is the strength of my signals?

QSB - Are my signals getting weaker? More!

QSD - Is my keying defective?

QSG - Shall I send (number) messages at a time?

QSK - Can you hear me in between your signals and may I break in? More!

QSL - Can you acknowledge receipt? More!

QSLL - I will QSL on receipt of your QSL card. More!

QSM - Shall I repeat the last message I sent to you?

QSN - Did you hear my transmissions on (frequency)?

QSO - Can you communicate with me? More!

QSP - Will you relay to (station)?

QST - General call preceding a message addressed to all Amateurs. More!

QSU - Shall I send or reply on this frequency?

QSW - Will you send on this frequency?

QSX - Will you listen on (frequency)?

QSY - Shall I change transmission to another frequency?

QSZ - Shall I send each word or group more than once?

QTA - Shall I cancel message (number)?

QTB - Do you agree with my word count?

QTC - How many messages do you have to send?

QTH - What is your location?

QTR - What is the correct time?

U.S. Amateur Bands

February 23, 2007   Extra (E), Advanced (A), General (G), Technician (T), Technician Plus (T+), Novice (N)

160 Meters
E, A, G   1.8 – 2.0 MHz CW, Phone, Image, RTTY/Data

80 Meters
E   3.500 - 3.600 MHz CW, RTTY/Data   3.600 - 4.0 MHz CW, Phone, Image
A   3.525 – 3.600 CW, RTTY/Data   3.700 – 4.0 MHz CW, Phone, Image
G   3.525 – 3.600 CW, RTTY/Data   3.800 – 4.0 MHz CW, Phone, Image
N, T+   3.525 – 3.600 CW only

60 Meters (USB phone only to five discrete 2.8-kHz-wide channels, 50 W ERP max. power, tune 1.5 kHz lower)
E, A, G   5.332 (5.3305), 5.348 (5.3465), 5.368 (5.3665), 5.373 (5.3715), 5.405 (5.4935) MHz.

40 Meters
E   7.000 – 7.125 MHz CW, RTTY, Data   7.125 – 7.300 MHz CW, Phone, Image
A   7.025 – 7.125 CW, RTTY/Data   7.125 – 7.300 MHz CW, Phone, Image
G   7.025 – 7.125 CW, RTTY/Data   7.175 – 7.300 MHz CW, Phone, Image
N, T+   7.025 – 7.125 CW only

30 Meters
E, A, G   10.100 – 10.150 MHz CW, Image, 200 watt maximum.

20 Meters
E   14.000 – 14.150 MHz CW, RTTY, Data   14.150 – 14.350 MHz CW, Phone, Image
A   14.025 – 14.150 CW, RTTY, Data   14.175 – 14.350 MHz CW, Phone, Image
G   14.025 – 14.150 CW, RTTY/Data   14.225 – 14.350 MHz CW, Phone, Image

17 Meters
E, A, G   18.068 – 18.110 MHz CW, RTTY/Data   18.110 – 18.168 MHz CW, Phone, Image

15 Meters
E   21.000– 21.200 MHz CW, RTTY/Data   21.200 – 21.450 MHz CW, Phone, Image
A   21.025 – 21.200 CW, RTTY/Data   21.225 – 21.450 MHz CW, Phone, Image
G   21.025 – 21.200 CW, RTTY/Data   21.275 – 21.450 MHz CW, Phone, Image
N, T+   21.025 – 21.200 CW only

12 Meters
E, A, G   24.890 – 24.930 MHz CW, RTTY, Data   24.930 – 24.990 MHz CW, Phone, Image

10 Meters
E, A, G   28.000 – 28.300 MHz CW, RTTY/Data   28.300 – 29.700 MHz CW, Phone, Image
N, T+   28.000 – 28.300 MHz CW, RTTY/Data   28.300 – 28.500 MHz Phone, CW, 200 watts PEP max

6 Meters
E, A, G, T, T+   50.0 – 50.1 MHz CW only   50.1 – 54.0 MHz CW. Phone, Image, MCW, RTTY/Data

2 Meters
E, A, G, T, T+   144.0 – 144.1 MHz CW only   144.1 – 148.0 MHz CW, Phone, Image, MCW, RTTY/Data

1.25 Meters
E, A, G, T, T+, Novice limited to 25 Watts   222 – 225 MHz CW, Phone, Image, MCW, RTTY/Data

70 Centimeters
E, A, G, T, T+   420 – 450 MHz CW, Phone, Image, MCW, RTTY/Data

23 Centimeters
E, A, G, T, T+   1240 – 1300 MHz CW, Phone, Image, MCW, RTTY/Data
Novice limited to 5 watts   1270 – 1295 MHz CW, Phone, Image, MCW, RTTY/Data

Higher Frequencies
All modes and licensees (except Novices) are authorized on the following bands [FCC Rules, Part 97.301(a)]:
2300-2310 MHz
2390-2450 MHz
3300-3500 MHz
5650-5925 MHz
10.0-10.5 GHz
24.0-24.25 GHz
47.0-47.2 GHz
75.5-81.0 GHz*
119.98-120.02 GHz
142-149 GHz
241-250 GHz
All above 300 GHz

* Amateur operation at 76-77 GHz has been suspended till the FCC can determine that interference will not be caused to vehicle radar systems

Amateur Radio Call Sign Numerical Prefixes

Region 1
- Maine (ME)
- New Hampshire (NH)
- Vermont (VT)
- Massachusetts (MA)
- Rhode Island (RI)
- Connecticut (CT)

Region 2
- New York (NY)
- New Jersey (NJ)

Region 3
- Pennsylvania (PA)
- Delaware (DE)
- Maryland (MD)

Region 4
- Kentucky (KY)
- Virginia (VA)
- Tennessee (TN)
- North Carolina (NC)
- South Carolina (SC)
- Alabama (AL)
- Georgia (GA)
- Florida (FL)

Region 5
- Texas (TX)
- New Mexico (NM)
- Oklahoma (OK)
- Arkansas (AR)
- Louisiana (LA)
- Mississippi (MS)

Region 6
- California (CA)

Region 7
- Washington (WA)
- Oregon (OR)
- Idaho (ID)
- Montana (MT)
- Wyoming (WY)
- Nevada (NV)
- Utah (UT)
- Arizona (AZ)

Region 8
- Michigan (MI)
- Ohio (OH)
- West Virginia (WV)

Region 9
- Wisconsin (WI)
- Illinois (IL)
- Indiana (IN)

Region 0
- North Dakota (ND)
- South Dakota (SD)
- Minnesota (MN)
- Nebraska (NE)
- Iowa (IA)
- Colorado (CO)
- Kansas (KS)
- Missouri (MO)

* Additional U.S. Prefixes: KL7 - Alaska (AK), KH6 - Hawaii (HI)

FCC Regional Call sign Groups

A unique call sign is assigned to each amateur station during the processing of its license application. The station is reassigned this same call sign upon renewal or modification unless application for a change is made. Each call sign has a one letter prefix (K, N, W), or a two letter prefix (AA-AL, KA-KZ, NA-NZ, WA-WZ), and a one, two, or three letter suffix separated by a numeral (0-9) indicating the geographic region. When the call signs in any regional-group list are exhausted, the selection is made from the next lower group. The groups are:

Group A: Primary stations licensed to Amateur Extra Class operators. Regions 1 through 10 - prefix is the letter K, N, or W, and a two letter suffix; or a two letter prefix with first letter A, N, K, or W, and one letter suffix; or two letter prefix with first letter A, and two letter suffix.

Group B: Primary stations licensed to Advanced Class operators. Regions 1 through 10 - prefix is two letters with the first letter K, N, or W, and a two letter suffix.

Group C: Primary stations licensed to General, Technician, and Technician Plus Class operators. Regions 1 through 10 - prefix is the letter K, N, or W, and a three letter suffix.

Group D: Primary stations licensed to Novice Class operators, and for club and military recreation stations. Regions 1 through 10 - prefix is two letters with first letter K or W, and three letter suffix.

Signal Reporting

Characteristics	Readability	Signal Strength
1	Unreadable	Faint signals, barely perceptible
2	Barely readable, occasional words distinguishable	Very weak signal
3	Readable with considerable difficulty	Weak signal
4	Readable with practically no difficulty	Fair signal
5	Perfectly readable	Fairly good signal
6		Good signal
7		Moderately strong signal
8		Strong signal
9		Extremely strong signal

A signal report "5 by 9" would indicate that the readability is "5" (perfectly readable) and the signal strength is "9" (strong signal)

Communications Net Procedure

Frequency, Offset, PL Tone

NET INTRODUCTION PROCEDURE
CQ, CQ, CQ, All Amateur Radio Operators. Does anyone need the repeater for emergency or priority traffic? (drop)

Calling together the (_____Organization) Net. This is (_____ your call sign). I will be your net control station for this session of the (_____Organization) Communications Net for (_____ date). This net is meeting through the facilities of the (_____ station) repeater. (drop)

The purpose of this net is to relay information for the (_____Organization). We welcome and encourage all amateur operators to check into this net if you have information or need help even if you are not a member of the (_____Organization). (drop)

Please remember that this is a directed net. Please do not break into the net without direction from net control. If you do have an emergency during the net please state the word "emergency" followed by your call and the frequency will be turned over to you for the duration of the emergency. If you have information for the net please use the word "info" followed by your call. If you have a question for the net please use the word "query" followed by your call. (drop)

When checking into this net please use the hesitation method. The procedure is as follows: say "THIS IS ...". Release your push to talk switch to see if you are doubling with anyone. If you do not hear anyone else talking then proceed to give your call sign phonetically, using the Standard ITU phonetic alphabet, your name and location. Also state if you have any information dealing with emergency communications for the net, then stand by and wait for net control to acknowledge you before doing anything else within the net. (drop)

Please do not leave this net unless you first inform net control and receive permission to do so. I will now stand by for anyone with EMERGENCY or PRIORITY TRAFFIC ONLY. (drop)

I will now stand by for the regular net check-in by townships. Please remember to give your call slowly and phonetically along with your name and location. Also please remember to state if you have anything for the net. (drop)

Order of check-ins:
LDOC
(location) Shelter
(location) Shelter
(location) Shelter
(location) Shelter
Non Shelter Operators

(ACKNOWLEDGE ALL STATIONS THAT CHECK INTO NET. STOP THE NET AND ASK THE NET TO STAND BY WHEN YOU NEED TO, SO YOU CAN WRITE DOWN EVERYONE CHECKING IN. WHEN NO FURTHER STATIONS CHECK INTO THE NET THEN PROCEED WITH THE NET. REMEMBER TO ASK FOR NEW CHECK-INS OR IF ANYONE HAS ANYTHING FOR THE NET EVERY FEW MINUTES [APPROX. 5 - 10]).

NET OPERATION PROCEDURE
LIST ANY BULLETINS OR PRIORITY MESSAGES YOU MIGHT HAVE FOR THE NET.
POLL EACH SHELTER TO SEE IF THEY HAVE ANY INFORMATION FOR NET.
HAVE EACH INDIVIDUAL STATION THAT STATED THEY INFO FOR THE NET LIST WHAT THEY HAVE AND PASS IT.
HANDLE ANY EDUCATIONAL INFORMATION TO BE COVERED DURING THIS NET.
** REMEMBER TO HAVE A PAUSE TO SEE ANYONE THAT COMES INTO THE NET LATE WISHES TO CHECK-IN.

NET CONCLUSION PROCEDURE
(CALL FOR ANY FINAL STATIONS WISHING TO CHECK INTO THE NET AND CHECK TO SEE IF THERE ARE ANY FURTHER COMMENTS OR QUESTIONS. IF THERE IS NO FURTHER ACTIVITY PROCEED TO CLOSE THE NET.

This is (_____ your call sign), (_____Organization) net control station for this session of the (_____Organization) Communications Net thanking all of the stations that checked in today for participating and supporting the net. I also want to thank those stations that did not check in for standing down while the net was in operation. Your help in maintaining this net is greatly appreciated by one and all (drop).

The (_____Organization) wishes to again thank (_____ station) for the use of his repeater for the operation of this net. All stations may now secure. This net is now secured at (_____ local time). This is (_____ your call sign) returning the repeater to its normal amateur operation. Good evening everyone (drop)

CTCSS (PL) Tone Frequencies

The purpose of CTCSS is to reduce co-channel interference during band openings. CTCSS repeaters would respond only to signals having the CTCSS tone required for that repeater. These repeaters would not respond to distant weak signals on their inputs and would not repeat those signals. Listed are the standard Electronic Industries Association (EIA) frequency codes, in hertz, along with their Motorola alphanumeric designators.

67.0 – XZ
69.3 – WZ
71.9 – XA
74.4 – WA
77.0 – XB
79.7 – WB
82.5 – YZ
85.4 – YA
88.5 – YB

91.5 – ZZ
94.8 – ZA
97.4 – ZB

100.0 - 1Z
103.5 - 1A
107.2 - 1B

110.9 - 2Z
114.8 - 2A
118.8 - 2B

123.0 - 3Z
127.3 - 3A
131.8 - 3B

136.5 - 4Z
141.3 - 4A
146.2 - 4B

151.4 - 5Z
156.7 - 5A
162.2 - 5B

167.9 - 6Z
173.8 - 6A
179.9 - 6B

186.2 - 7Z
192.8 - 7A

203.5 - M1
203.5 - M1

206.5 - 8Z
206.5 - 8Z

210.7 - M2
218.1 - M3
225.7 - M4

229.1 - 9Z

233.6 - M5
241.8 - M6
250.3 - M7

254.1 - 0Z

Packet Radio

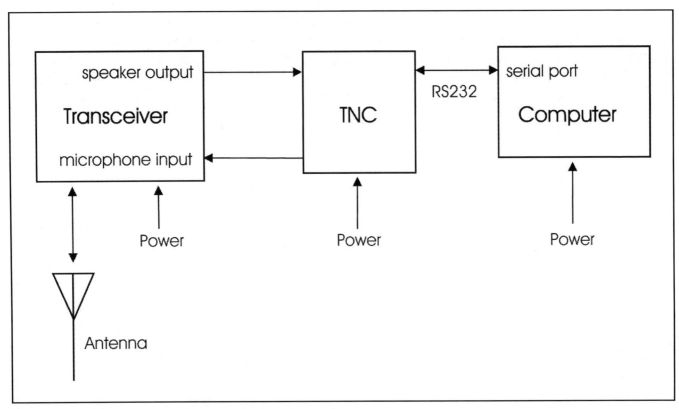

Introduction

The equipment needed to get on the air is:

- Radio transceiver (without the microphone).
- Computer or terminal (probably none of the latter exist any longer).
- Terminal node controller (TNC).
- Appropriate interconnection cables.

There is packet activity on HF, but VHF is the best place to begin operating Packet Radio. The Terminal Node Controller (TNC). The TNC contains special firmware especially designed for Packet Radio. This "firmware" converts computer data into "packets" of digital information that can be sent (error free, via Packet Radio nodes) across the Packet Radio network. Some radios, such as the Kenwood TMD700A have a built-in TNC.

Keyboarding with friends over long distances is easy when using a Packet Radio Network. Very little power is needed, as the nodes along the network perform as "Automatic Routing" devices. The nodes do the " Automatic Routing " function automatically. The Packet Radio user only has to establish the connection. The rest is handled by the nodes.

The TNC

A "Terminal Node Controller" (TNC) is similar to the modem used in computers. The TNC is used to interface the terminal or computer into the "RF" or radio (wireless) medium. Inside the TNC most manufacturers have added some internal firmware called a "PAD." The pad or "Packet assembler/dissembler" captures incoming and out-going data and assembles it into "packets" of data that can be sent to and from a data radio or transceiver.

The PAD also enables the Push-To-Talk (PTT) circuits of the radio transceiver. When the enter key of the computer keyboard is pressed, the typed in data is sent out over the air to the target station or a nearby "store-and-forward" device known as a "node."

Incoming (received) data from the transceiver is also converted within the PAD, from Packets of data into a stream of usable data and sent to the TNC/modem. The data stream is then sent to the serial comport of the computer for display on the screen, or manipulated by a resident terminal program into on-screen text, pictures, or save-to-disk processing.

Usually an RS232 DB25 male to DB9 female cable is used to connect the TNC to the computer. The RS232 cable can be purchased. The TNC can connect to the radio either via the microphone connector or a separate data connector, which is usually a DIN connector. The microphone or data connector usually must be wired according to the manufacturers specifications.

There are TNC's and there are TNC's. A low cost TNC will work fine. However, you will need to put many control codes into it or use a software program that can add these codes. The more expensive TNC's have many control codes built-in. With built-in control codes, you can use simple software programs like "hyper Terminal," which is free with windows. In my station, I'm using the Kantronics KPC3 Plus and Hyper Terminal.

Transmit Level Adjustment

The transceiver drive should be about 3 to 3.5 kHz. of deviation. This can be measured using a deviation meter if available. If not, check the manual of the transceiver for the packet audio input voltage (for the ICOM 207H it is 400 mV). Next, check the manual of the TNC for the procedure to adjust the output audio voltage of the TNC. If the appropriate equipment is not available, some trial and error may work.

Antenna Height

If your attempting to work direct over long distances, without the use of a FlexNet Digi (Node) or a BBS, no antenna can be big enough nor high enough !

Using a well located Node or BBS is similar to voice operation through a repeater, wherein a handi talkie and rubber duck antenna is often sufficient.

Software

A terminal program is needed to control the data going to and from the TNC and radio. Software can range from a simple DOS or Windows program to more sophisticated software that has logging and other functions. Some manufacturers include software with their TNC's.

There are two communication speeds that are used in Packet Radio 1200 Baud or 9600 Baud. Set the computer serial port to the speed of the computer serial port, usually 9600 Baud. Set the packet system speed to 1200 or 9600 Baud. ABAUD refers to the computer to TNC (serial port), and HBAUD refers to the RADIO or ON-AIR Baud rate (data speed).

There are a number of packet software packages available costing close to $100. Personally, I don't like any of the ones I have tried. I like and use the "Hyper Terminal" program that is supplied with Microsoft Windows. Set Hyper Terminal for the com port you are using, which is usually COM1, the speed at which the com port works, which is usually 9600 bps, Data bits = 8, Parity = none, Stop bits = 1. After you complete the settings, save them as "Packet."

For those that prefer a more "user friendly" packet terminal for older PC's, DOS based "paKet" is excellent. Windows based WinPack (know in Europe as TPK) is also good. Any of these "shell" programs do greatly simplify the TNC commands that are detailed in the following paragraphs. They also enable you to multistream, such as remain connected to a DX Cluster while you check your mail and hold a keyboard to keyboard conversation with a friend!

Packet Operation

Switch the transceiver ON and turn the volume up a quarter turn or just above the "9:00 o'clock position." Make sure the squelch is not set too tight. The squelch should be set to a position where the transceiver is quiet. The squelch is set in a similar manner that you would use for voice operation. When first turned on, the TNC you may display garbled text on the screen. This is usually because the terminal to TNC baud rate is not set to the same parameters. Some TNC's will do a "search" mode to find the proper settings.

Perform a "control C" **[Ctrl C]** (press Ctrl and the letter C at the same time) to place the TNC into the command (cmd:) mode. This is where all commands are made to and from the TNC. Any command that is typed while in the "cmd: mode is received by the TNC as a direct order. These codes can vary with TNC's.

Once in the command mode, press the [Enter] key. Each time the [Enter] key is pressed a "cmd:" prompt should appear on the screen. This is an indication that the computer has control (command) of the TNC.

> **All commands must be followed by the [Enter] key.**
>
> The next step will be to set the station call sign into the TNC. At the cmd: prompt, type:
> **MY *(your call sign)***

Test the TNC to see if the station call sign is set into the TNC. To do so, type:
MY

The screen should display a response from the TNC with:
MYCALL *(your call sign)*

MYCALL NOCALL indicates that a call sign has never been set, or the internal memory battery has been disconnected or is dead.

To enter your call sign type:
MY *(your call sign)*

The TNC should respond with:
MYCALL *(your call sign)*

This indicates that the computer and TNC are communicating properly. If there is no response after typing MY, then try typing:
ECHO ON

The :cmd: should appear on the screen again, with a message similar to the following:
ECHO was OFF

If the computer is displaying double letters, (for example; MMYY CCAALLLL), this indicates that the ECHO command should be turned OFF. Type the following:
ECHO OFF

The TNC should respond with:
ECHO was ON

Below are some commands that should be made active:
ECHO ON (normal) or ECHO OFF (if double letters are displayed)
MONITOR ON
MCOM ON
MCON OFF (to display only packets addressed to you) or MCON ON (to display all packets)
MRPT ON

If the RS-232 interface cable is wired using the RTS, CTS, Txd, Rxd, and Signal Ground leads, then set the XFLO command OFF. If the RTS, and CTS signals were not used, then make sure the XFLO command is ON.

Note: TNC's have 3 modes of operation: Command, Converse and Transparent. You must remain aware of which mode the TNC is in at any current moment!

Command Mode

In the COMMAND mode, the TNC will interpret data received from the keyboard as a command to process data, not as data to transmit.

When you are in the command mode, the screen will display:
cmd:

Brief list of NEWUSER commands:
CONMODE CONVERS (TNC will automatically be placed into CONVERS mode after connection is
estabished
CONMODE TRANS (TNC will automatically be placed into TRANS mode)
CONNECT or **C** (connects to another station)
CONVERS (to enter convers mode)
DAYTIME (to read the time and date)
DAYTIME yymmddhhmm[ss] (to enter the time and date)
DIGIPEAT ON (turns digipeat on)
DIGIPEAT OFF (turns digipeat off)
DISCONNECT or **D** (disconnect from another station)
DWAIT n (n=0-255) (10 times n in milliseconds) (delay used to avoid collisions betweendigipeated packets)
ECHO ON (character received from the keyboard are echoed back to the screen)
ECHO OFF
HELP (for most TNC's will generate a list of commands)
INTFACE NEWUSER (for most TNC's will enter standard terminal mode with a limited command set
INTFACE TERMINAL (for most TNC's will enter terminal mode with full command set
MCOM ON (monitors all packets being transmitted)
MCOM OFF
MCON ON (will display all packets received)
MCON OFF (will display only packets addressed to you)
MHEARD SHORT (short list of stations heard * indicates digipeating)
MHEARD LONG (long list of stations heard)
MHEARD CLEAR (clear the list of stations heard)
MONITOR ON (unconnected packets will be seen. Also acts as a master control for MALL, MCOM, MCON, MRESP, MRPT)
MONITOR OFF
MRESP ON (monitors packets including AX.25)
MRESP OFF
MRPT ON (entire digipeat list is displayed)
MRPT OFF
MYALILAS xxxxxx-n (n= 0-15) (sets TNC to an alias call sign for digipeating)
MYCALL xxxxxx-n (n=0-15) (sets TNC for you call with the optional supplementation Station Identifier (SSID)
NOMODE ON (TNC does not change modes after a connection is established)

NOMODE OFF (TNC will change to whatever mode is established in CONMODE after a connection is established)
TXDELAY n (n=0-255) (delays transmit to give your radio enough time to reach full power, set delay to 10 times n in milliseconds) (300 ms is commonly used)
UNPROTO CALL VIA W2ABC,W2CDF,W2EFG (max 8 call signs)

Convers (Conversation) Mode

In the CONVERS mode, the TNC will interpret data received from the keyboard as data to be transmitted. Most TNC's will automatically switch to the CONVERS mode after a connection has been established. When you are in the COMMAND mode, you can switch to the CONVERS mode by giving the command:
CONVERS or **K**

If you are in CONVERS mode and want to switch to COMMAND mode, type:
[Ctrl] C

Trans (Transparent) Mode

A second method for transmitting data, called TRANS mode, is to instruct the TNC to ignore "control characters," such as "backspace," and transmit every character as data. For many TNC's TRANS mode is a TERMINAL mode not a NEWUSER mode.

If you are in TRANS mode and want to switch to COMMAND mode, type:
[Ctrl] C three times with a pause of less than second between entries

Monitoring or Calling CQ

If you turn the MONITOR command on, you will see other packet stations you your screen. You will see two call signs at the beginning of each packet separated by a ">" The first station is the station that is sending the packet. The second is the station receiving the packet.

To call CQ, you must be in the CONVERS mode, so that the data received from the keyboard will be interpreted as data to be transmitted.

To enter the CONVERS mode, type:
CONVERS or **K**

Anything you type at this point, will be transmitted.

Example:
W2XYZ CQ CQ CQ

If a station wants to connect to you, they will type the CONNECT W2XYZ command

To return to the COMMAND mode, type:

[Ctrl] C

Packet Direct

The most common frequency for packet communications is 145.010 mhz at 1200 Baud.

Begin in the command mode:
[Ctrl] C

Enter your call sign into the TNC
MY *(your call sign)*

Test that the TNC has received you call sign:
MY

The screen should display a response from the TNC with:
MYCALL *(your call sign)*

To connect directly to W2XYZ, assuming you both have a direct path:
CONNECT W2XYZ or C W2XYZ

If the TNC receives an acknowledgement of connection it will display:
*** CONNECTED TO W2XYZ

Once connected, the TNC should automatically switch to conversation mode (CONVERS). You can type in text, then press enter to send. You should automatically receive text from the station you are connected to.

When you have completed your conversation, you need to get back to COMMAND mode to sign off.

To get back to COMMAND mode, type:
[Ctrl] C

To disconnect, type:
DISCONNECT or D

The TNC should respond with:
*** DISCONNECTED

Connecting through a Node

1. To connect to W2XYZ thru node WA2PNU, assuming both stations are listening to WA2PNU node:
   C W2XYZ V WA2PNU

2. To connect to W2XYZ thru node NY2LI, assuming your listening to WA2PNU node and W2XYZ is listening to NY2LI node:
   C W2XYZ V WA2PNU NY2LI

3. To connect to W2XYZ thru distant node K2JFK (Clay NY), assuming your listening to WA2PNU and W2XYZ is in Clay NY listening to node K2JFK:
   C W2XYZ V WA2PNU K2JFK

4. Above examples are from a disconnected state. You can connect first to your local node,
   C WA2PNU, and then the WA2PNU call can be deleted from the previous examples, such as:
   1) **C W2XYZ**   2) **C W2XYZ V NY2LI**   3) **C W2XYZ V K2JFK**

5. To find out what node W2XYZ is monitoring, on an ARRL section by section basis, you connect to any node within that section, and then do a "find".
   Example:
   F W2XYZ

6. "A" command on any FlexNet will give a manually built list of nodes with geographic locations. Some sites have newer updates than others. "D" gives the machine made <D>estination list, showing callsigns, SSID range, and "round trip times" of other nodes. The D list will always be up to date! Nodes with RTT's under 1000 should be easily connected to, over about 1000 means the path may be dropping out due to propagation conditions.

   Example (with your radio set to 145.07 MHz):
   C W2PNU

   *** CONNECTED to W2PNU
   PC/FlexNet V3.3g West Hills, LI, NY, USA
   1200 baud 145.07   9600 baud 145.59
   <C>onnect <D>estinations <F>ind <H>elp <I>nfo <MH>eard <P>orts <Q>uit <U>sers
   <A> for Callsign vs Location Table <M>ail will connect to the nearest BBS

Disconnecting from a node

Q for <Q>uit on FlexNet nodes, remember command is B for <B>ye on FBB BBS's!

Bulletin Board Servers

Bulletin Boards are a "Store and Forward" device. Once you post you message or bulletin, the
server stores it and then passes it on the neighboring BBS's.

Typically, Personal messages and NTS traffic are forwarded instantly, wherein Bulletins may be
delayed until off peak hours as not the tie up the network.

Connecting to a BBS

BBS's usually are co-located with a Node. But not all Nodes have a BBS.

The nodes usually have the path a BBS programmed in. The M command will connect you to the BBS (<M>ailbox).

Users can either connect to their nearest Node, and then connect onward to the BBS, or just connect directly to the BBS itself.

Example:
Starting on 144.99 MHz, **C WA2PNU**, then **M**.
Or starting on 144.99 MHz, just **C WA2PNU-4**
But starting on 145.07 MHz, **C W2XYZ**, and then **M**, will also connect you to WA2PNU-4 BBS!

Example:
C WA2PNU
*** CONNECTED to WA2PNU
PC/FlexNet V3.3g West Hills, LI, NY, USA
1200 baud 145.07 9600 baud 145.59
<C>onnect <D>estinations <F>ind <H>elp <I>nfo <MH>eard <P>orts <Q>uit <U>sers
<A> for Callsign vs Location Table   <M>ail will connect to the nearest BBS

M (command is short for <M>ailbox)
link setup...
*** connected to WA2PNU-4
[FBB-7.00g-AB1FHMRX$]
WA2PNU BBS, QTH FN30HU.
Hello John, you are now on channel 1.
Here are 864 active messages, 229214 is last message and 228445 is the last you have listed.
Assigned channels:
Ch. 1 (FLEX) : WB2LUA-0 - MSP W2XYZ
on 01/04/02 10:43
via : WA2PNU-0 W2XYZ-2

Ch. 2 (FLEX) : KB2VLX-4 - Mon 01/04/02 10:42
via : WA2PNU-0
WA2PNU BBS (H for help) >

Abbreviated list of available FBB BBS commands:

A	Abort - Abort listing.
B	Bye - Log off the BBS.
H	Help - Help.
K	Kill - Kill messages.
L	List - List messages.
M	Make - Copy a message to a file.
N	Name - Change your name.
NQ	Your Home QTH, "City, ST"
NZ	Zip - State your zip-code.
NH	homeBBS - Type your home-BBS.
O	Option - Select options (paging, language, list/read personal etc).
R	Read - Read messages.
S	Send - Send messages.
X	Expert – Shortens the command prompt line

Supplemental Identifier (SSID)

By adding -0 to -15 after your call sign, the operator can use their call sign 16 times.

Types of messages

There are three basic types of messages, Personal (P), Bulletin (B), and NTS Traffic (T).

Personal messages are from one user to a second user, while Bulletins are from one user to a group of users in a designated area.

The difference being that (P) messages route only to the intended recipients home BBS while (B) bulletins flood every BBS in the designated area.

Example:
A bulletin may be sent to TRIBBS, NEBBS, NYBBS, CTBBS, USBBS or WW, meaning respectively the Tri-State Metro area, New England BBS's, New York BBS's, Connecticut BBS's, BBS's throughout the USA, or BBS's throughout the entire World.

NTS traffic is third party mail, forwarded via Postal ZIP codes. It is routed to the nearest BBS's to the third parties street address.

Listing messages

The following commands are used to list messages

L lists every message on the BBS, back to the marker of the previous last message you listed.
Beware that this command will list the last couple weeks or more of messages (think in terms of several thousand) the first time you long on to a BBS as a new first time user.

LL ## lists the last ## number of messages. Again this command will list (P), (B) and (T) messages.

LM Will list (P) messages only addressed to you

LB will list only (B) bulletins

LT will list only (T) NTS traffic

LS (subject) will list messages containing the subject in message title
Example:
LS DX will list very message with "DX" in the title.

Receiving a Message

R ### where ### is the message number will give you the text of that message, be it (P) (B) or (T).

With (P) mail, it is polite to kill any message you have received and read with either the K ### or KM commands.

K ### will kill one message, as specified by the number ###

KM will kill all messages addressed to you.

In the case of NTS traffic, once you have delivered the message, it is proper to log back onto the BBS and kill that message.

Receiving Private Mail

RP (Receive Private) or **R ###**, where ### is the message number

KM (Kill Message) to delete the current message

To Send a Reply:
SR (Send Reply) or **SR ###**, where ### is the message number

Sending Private Mail

SP K2HAM (Send Private)
Routing (from WP) to K2JFK.#CNY.NY.USA.NOAM.
{White Paper (WP) server transparently learns users home BBS address's from traffic passing through each BBS, and in turn automatically shares this information between BBS's}

Enter the title for this message to K2HAM

Text Message
Enter the text for the message, end with Ctrl-Z or /EX on a blank line)
Hello test, 1 2 3 4 5 6 7 8 9 10
73 de Jose
/ex   {Always end with /EX on a new line}

Mid: 32607_WA2PNU Size: 85 bytes
WA2PNU BBS (H for help) >

An (P) mail address must follow this format:
K2HAM@K2JFK.#CNY.NY.USA.NOAM
^addresses call sign
^addresses home BBS
^supplemental geographic info (often ARRL Section)
^State
^Country
^Continent

Sending a Bulletin

SB RACES @ TRIBBS (Send Bulletin)
(this bulletin will flood all the BBS's in the Tri State area)

Enter the title for this bulletin:
Monday Nights Test Message

Enter the text for the message, end with Ctrl-Z or /EX on a blank line)
**Hello test, 1 2 3 4 5 6 7 8 9 10 from the EOC in Huntington, LI, NY.
73 de Jose, RACES Officer, Huntington
/ex**   {Always end with /EX on a new line}

Mid: 32607_WA2PNU Size: 185 bytes
WA2PNU BBS (H for help) >

Sending NTS TRAFFIC

When the message is ready to be entered into your local BBS, you must use the **ST** command,
which means "Send Traffic", followed by the zip code of the destination city, then @ NTS followed
by the two letter state abbreviation. The form used is:
ST ZIPCODE @ NTSxx (send NTS traffic)
Example:
A message being sent to Boston, MA 02109 would be entered as follows:
ST 02109 @ NTSMA

Enter the title for this bulletin:
Test Message

Enter the text for the message, end with Ctrl-Z or /EX on a blank line)
Hello test, 1 2 3 4 5 6 7 8 9 10 from the EOC in Huntington, LI, NY.
73 de Jose, RACES Officer, Huntington
/ex   {Always end with /EX on a new line}

Mid: 32607_WA2PNU Size: 185 bytes
WA2PNU BBS (H for help) >

To Send a Reply:
SR (Send Reply) or **SR ###**, where ### is the message number

Disconnecting from the BBS

B for <B>ye on FBB BBS's, remember command is Q for <Q>uit on FlexNet Nodes!

You have been connected 5mn 14s - Computer-time: 9s
Bye, John, and welcome back.
*** reconnected to WB2CIK

APRS (Automatic Packet Reporting Service)

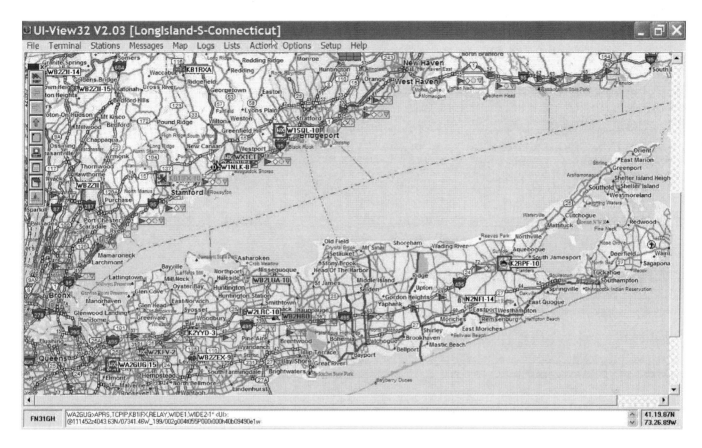

Automatic Packet Reporting Service is a specialized form of packet operating on a single frequency, which is 144.390 MHz in the USA. If one reaches an IGate, the position information transmission will enter the internet. It can be seen using: Google Maps for a region (example Long Island, NY http://aprs.fi/?lat=40.8287&lng=-73.3962&timerange=86400), and Findu (example: http://www.findu.com/cgi-bin/find.cgi?call=wb2lua-3&terra=4). Mobiles often use GPS receivers to automatically send beacons. Fixed stations mostly use their fixed location coordinates. UI-View 32 is the PC-based mapping software that is often used. Mobile radios such as the Kenwood D710a have built in TNC's and only require a GPS receiver to beacon on APRS. Only an internet connection and computer are required to see the beacon locations. APRS was developed by Bob Bruninga, WB4APR.

During emergencies, APRS can be very valuable. It can be monitored at an emergency operations center to track emergency response teams, mobiles, etc. Responders in the field can also see their own location.

The Global Positioning System is maintained by the U.S. Department of Defense and consists of 25 satellites in orbit around the earth. Positioning information is determined by a small receiver which measures the time in micro-seconds that it takes to receive the broadcast from between 1-12 satellites. By receiving the signal from at least four satellites, position information down to about 10 meters can be determined. Altitude information can also be obtained from the system. In mobile situations you can determine speed and direction.

SSID's

-0 Your primary station usually fixed and message capable
-1 generic additional station, digi, mobile, wx, etc
-2 generic additional station, digi, mobile, wx, etc
-3 generic additional station, digi, mobile, wx, etc
-4 generic additional station, digi, mobile, wx, etc
-5 Other networks (Dstar, Iphones, Androids, Blackberry's etc)
-6 Special activity, Satellite ops, camping or 6 meters, etc
-7 walkie talkies, HT's or other human portable
-8 boats, sailboats, RV's or second main mobile
-9 Primary Mobile (usually message capable)
-10 internet, Igates, echolink, winlink, AVRS, APRN, etc
-11 balloons, aircraft, spacecraft, etc
-12 APRStt, DTMF, RFID, devices, one-way trackers*, etc
-13 Weather stations
-14 Truckers or generally full time drivers
-15 generic additional station, digi, mobile, wx, etc

New nN Paradigm

1. RELAY, WIDE, TRACE, TRACEN-N and SS are obsolete.
2. Use WIDE2-2 for fixed stations (3-3 is ok for areas far from any city or mountains)
3. WIDE1-1,WIDE2-1 for mobiles (WIDE1-1, WIDE2-2 is ok for ares far from any city or mountains.
4. Use WIDE1-1,SSn-N for selected non-routine State or Section nets or when humans are present for a large area emergent needs. (will not work reliably during transition period)
5. Use DIGI1,DIGI2,DIGI3... for point to point communications (but realize success becomes vanishingly small beyond 2 hops)
WARNINGS:
    A. Never use WIDE1-1 beyond the first hop
    B. Never use anything other than WIDEn-N on a Balloon for aircraft. (N=2 should work well)

UI-View Software

Download UI-View V2.03 from the UI-View website http://www.ui-view.org. There are also maps and other UI-View Add-Ons and information at this site.

UI-View requires a Registration Key which you can obtain from www.apritch.myby.co.uk Note: Registration is only available to Licensed Amateurs.

To install UI-View, run 32full203.exe and follow the prompts.

Next install maps for the local area. Download maps from the UI View website or make them from Street Atlas 9. Unzip the map files and copy them into the UI-View Maps directory C:\Program Files\Peek Systems\UI-View32\Maps

Comms Setup

From the Setup Menu, select Comms setup and select the type of TNC, Comm Port, etc. as seen below. The Kantronics KPC3+ works very well for this application and for Winlink 2000.

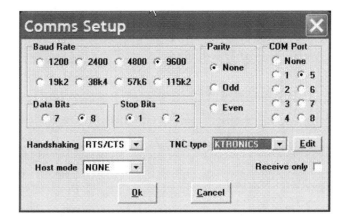

Station Setup

From the Setup Menu, select Station Setup and configure as seen below.

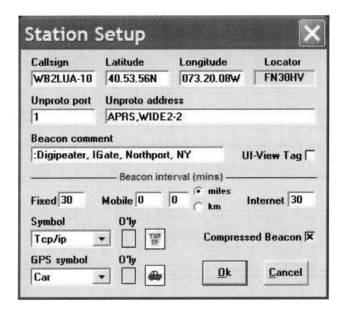

APRS Server Setup

From the Setup Menu, select APRS Server and configure as seen below.

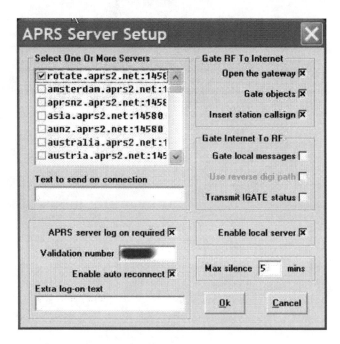

APRS Servers:
1. From the File menu, Click on "Download APRS Server List"
2. Type in: www.aprs2.net/APRServe2.txt
3. Click on the Download button. A list of APRS servers will be displayed.
    Some Regional Servers are:
        maine.aprs2.net:14580
        midwest.aprs2.net:14580
        northwest.aprs2.net:14580
        southwest.aprs2.net:14580
        or rotate between Tier2 Servers: rotate.aprs2.net:14580

        Filter for extra long-on texts may be added as: filter m/100

4. Check the box for the server selected and enter the validation number that came with your registration document.

To Add a url:
1. Press the Insert key
2. Type in the url
3. Press enter

To Remove a url:
1. Click on the entry and press the delete key

Miscellaneous Setup

From the Setup Menu, select Miscellaneous and configure as seen below.

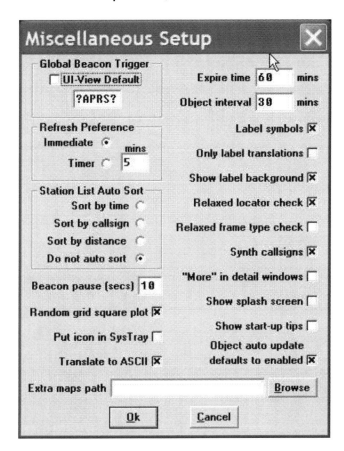

APRS Compatibility

From the Setup Menu, select APRS Compatibility and configure as seen below.

Messaging Setup

From the Message Menu, select Setup at the Message Retries dialog box. Set the *Retry interval, Try, Retry on heard* and *Expire after* as shown below.

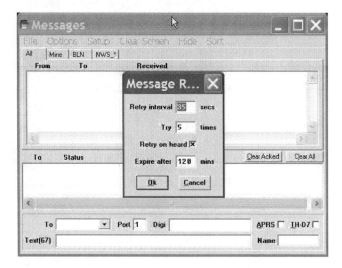

Connecting to APRS-IS

From the Files Menu, select Schedule Editor, then New. Then, select *APRSERVE_CONNECT* from the pull down menu in the *Command* box. Type: +1 in the *Time* box. Then click Yes. Then exit. This will connect UI-View Client to APRS-IS 1 minute after it has started.

Or one can click on Connect to APRS Server from the Action menu.

Phase Shift Keying (PSK) Radio

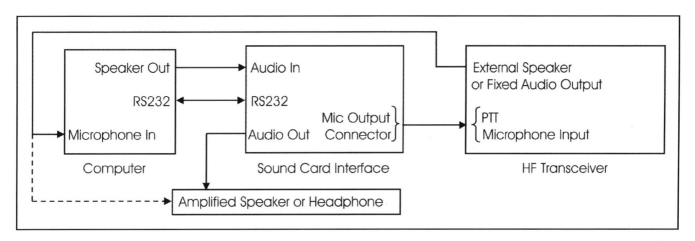

PSK31 (Phase Shift Keying) is a data mode that uses a personal computer and sound card to communicate. Packet is primarily designed for communications between two people or bulletin board. PSK31 is designed for multiple users like the voice nets. Anyone listening can see what everyone else is sending. PSK31 had many advantages over other modalities, such as requiring lower transmit power and more immunity from noise and interference (QRM). It uses an alphabet that has a text speed of 50 wpm. It does not require any handshaking with a second station. Roundtable communications are common in PSK31 mode. PSK31 was developed by Peter Martinez G3PLX.

The information in PSK31 is transmitted in patterns of reversed-polarities or 180-degree phase shifts. Phase modulation has several advantages over CW, which uses on-off keying. In a noisy or distorted propagation environment, the amplitude of CW will shift and vary much more than the phase of a signal.

The baud rate is 31.25 and the bandwidth is 31 Hz using narrow CW filters. The normal bandwidth of other modes is approximately 300-500 Hz. PSK31 can used with lower signal levels in a crowded digital band. PSK31 operates in a much narrower bandwidth than FSK (Frequency Shift Keying).

PSK63 is a variation of PSK31. It has a bandwidth of 63 Hz. and a speed of 100 wpm. RTTY has a speed of 60 wpm. PSK63 has improved polar path performance over PSK31.

The difference between a CW filter of 500 Hz and the bandwidth of PSK31 of 31 Hz (10*log(500/31) db = 12 db) is 12 db, which demonstrates that a CW transmitter must transmit 16 times more power than a PSK31 transmitter to achieve the same signal to noise ratio. Therefore, a PSK31 station can operate at 16 times less power than a CW station.

PSK creates a problem of key-clicks. The solution for eliminating key-clicks is to filter the output or to shape the envelope amplitude of each bit. The same problem of key-clicks may appear at the receiving end. PSK31 can eliminate this problem by filtering the receive signal or by shaping the envelope of the received bit. If a simple cosine wave is used at the receiver, a signal from one receive bit may be spread into the next bit. At the receive end, 4 bits are shaped at a time. The transmit and receive filters must be matched to each other. Over-driving the audio can create intermodulation products if it is not linear. So, it is important to not over-drive the audio.

BPSK (Binary PSK) mode that does not have forward error correction but is probably the most common mode on the bands. It can be identified by its two vertical lines in the "Vector" signal view window.

QPSK (Quadrature PSK) is another mode whereby instead of phase reversals (180 degree phase shifts), and additional pair of 90 and 270 degree phase-shifts are possible. It is like having two PSK (BPSK) transmitter on the same frequency, but, 90 degrees out of phase with each other. The result is twice the bit rate and 3 db less signal-to-noise ratio. QPSK mode has forward error correction but is a little harder to tune. It can be identified by its two vertical lines and two horizontal lines in the "Vector" signal view window. It is also sideband sensitive. Sometimes lower sideband is used.

PSK uses a personal computer and a 16 bit computer sound card. The audio output from the sound card is connected to the audio input of the transceiver with a 100:1 voltage divider to reduce the voltage from the sound card audio output to the transceiver audio input. Some interfaces use transformer isolation and some use opto-isolation. There a number of radio sound card interfaces available commercially. Most include a microphone connector, wire with microphone plug, bypass switches, a computer RS232 connector, and audio inputs and outputs.

Software

There are a number of software programs that can be used for PSK. This author used the highly recommended WinPSKse. It is freeware available at http://www.psk31.com along with various article and technical information on PSK31. WinPSKse is an adaptation of AE4JY's fine WinPSK program expertly crafted by Dave Knight, KA1DT. WinPSKse has the ability to display and read two PSK31 signals at the same time, in an easy to read and interpret presentation. This has a much improved user interface. For example, it has an amazing new simultaneous spectrum/ waterfall display.

Some PSK31 calling frequencies

BPSK primarily uses upper side band mode.
1,838.150 kHz
3,580.150 kHz to 3.620 kHz
7,035.150 kHz for region 1 and region 3, and 7080.15 for region 2 (the Americas)
10,142.150 kHz
14,070.150 kHz (Primary calling frequency)
18,100.150 kHz
21,080.150 kHz 150 (although most activity can be found 10 kHz lower)
24,920.150 kHz
28,120.150 kHz
145,550 kHz

Receive Audio Input Level

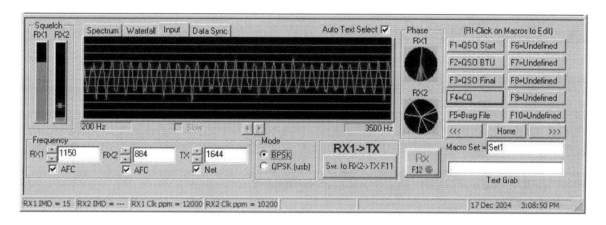

Tune the radio to a loud signal or carrier at 14,070.150 kHz. Display the soundcard's mixer program (or use the one that comes with Windows). Set the mixer's LINE IN setting to mid way. Adjust the volume control on the radio while viewing the INPUT signal display in the software program. The volume should be adjusted for a good signal level that is not too high and not too low.

Tuning in a PSK31 Signal

In the spectrum display view, look for peaks. Click the mouse on a peak to change the receive frequency marker position. If the display is not showing anything, adjust the soundcard Recorder mixer volume control or the volume control of the radio. Below is a typical PSK31 signal (RX1).

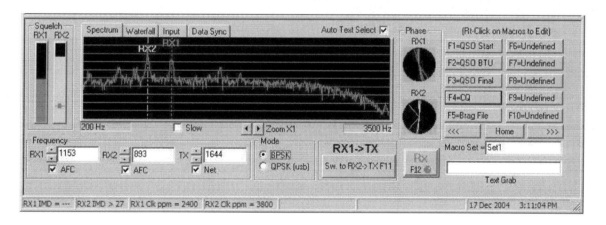

Transmit Audio Level Adjustment

Transmit level is more complicated to adjust than the receive level because the actual signal spectrum coming out of the transmitter cannot be seen at the transmitter. Adjust the mixer's VOLUME setting to adjust the transmit audio level The best method is to guess at a good level (mid way), then get a critical signal report over the air. The correct setting will vary from radio to radio. It is better to under-drive the radio until a clean signal is clean is obtained. In general never drive the transmitter anywhere near its rated power at first.

Operation Hints and Tips

The actual Transmit/Receive frequency is the USB radio dial frequency plus the audio frequency displayed in software. If using LSB, subtract the audio frequency from the radio dial setting. For example if the transceiver is in the USB mode and reads 14070.00 KHz and the audio frequency is 1500 Hz, then the actual transmit/receive frequency is 14071.50 KHz.

The TX and RX frequencies are limited between 200 and 3500 Hz. It is best to avoid the edges because transmitters may have some frequency limitations as well as some soundcards.

Don't send all text as UPPER CASE letters. PSK31 was designed to send the most commonly used letters such as 'e' and 't' much faster than letters such as 'z' that are used less frequently. Uppercase letters take much longer to send and slow down the transmission. Capitalize letters as needed. A common practice is to send callsigns in upper case.

Make sure the PC time and date are set correctly.

Try using the QPSK mode when conditions get rough. In many circumstances, using QPSK will greatly improve reception due to its error correcting capability.

Winlink 2000

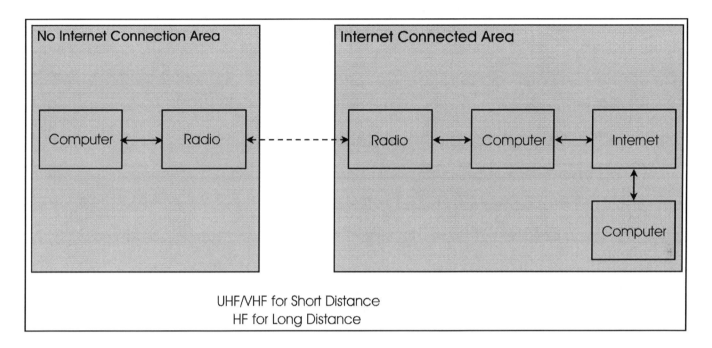

Winlink 2000 is a worldwide digital amateur radio message transfer system. It provides E-mail transfer with attachments, map & text-based position reporting, graphic & text-based weather bulletin services, and emergency communications by linking radios to the Internet.

The Winlink 2000 system is currently being utilized for emergency communications where local or regional communications are disrupted, including the loss of the Internet, and where accuracy of information is important.

Winlink 2000 can be used by any licensed Amateur radio operator. The operator logs into one of the participating network stations using the "AirMail" software. Currently, Winlink 2000 has a flow of over 150,000 messages monthly into 41 participating stations from 5100 + users. The Winlink 2000 user must have a General Class or higher license to use HF radio.

Winlink 2000 may be very useful for emergency communications using the Telpac with Paclink email-based VHF/UHF radio Packet for "last mile" communications coverage. Airmail is used for greater distances using the HF radio link to Winlink 2000 and the internet.

"Telpac" stands for TELnet-PACket Bridge and it allows the Winlink 2000 operator to use the VHF/UHF Packet mode with the B2F protocol. Telpac is used by the Packet nodes to interface with the end-user, who is using Paclink or Airmail. Paclink utilizes Outlook or Outlook Express to provide the end-user with a connection to the Winlink 2000 system by way of Telpac.

PACTOR is an HF (3 to 30 MHz.) radio teletype mode developed in Germany by Ulrich Strate (DF4KV) and Hans-Peter Helfert (DL6MAA) to improve on inefficient modes such as AMTOR/ SITOR and Packet-Radio (AX.25) in weak short wave conditions. PACTOR offers a much better error correction system, and a considerably higher data transfer rate, than AMTOR/SITOR and result in a protocol much more resistant to interference than Packet-Radio under poor propagation conditions. For the first time in amateur radio, online data compression is used to increase the effective transmission speed. Pactor I is capable of 200 bps. PACTOR II is capable of 800 bps. PACTOR III is capable of 3600 bps. To use PACTOR, a PACTOR TNC/Modem is used in place of a packet TNC and HF frequencies are used.

The dial (transmit) frequency is 1,500 KHz lower than the center frequency.

Current PACTOR nodes in the United States:

AB7AA - Bill in Waikiki Beach, Oahu, Hawaii, Scan Center Frequencies
3641.9, 7103.7(P3), 10142.7(P3), 14064.4, 14109.2(P3), 18104.9, 18106.2(P3)

AH6QK - Richard in Kaneohe, Oahu, Hawaii, Scan Center Frequencies:
7070.9, 10126.9, 14069.0, 14110(P3), 18101.9

KA6IQA - Tom in Rancho Santa Fe, California, Scan Center Frequencies:
7066.9, 7101.2(P3), 14112.4, 14104.2(P3)
18102.9, 18106.7(P3), (13:00 to 03:00 UTC)

KB6YNO, Hamilton, Massachusetts, Scan Center Frequencies:
7069.9, 10125.9, 14067.9, 14094.9 (P3), 18098.9

KF6NPC - Mike in Riverside County, CA., Scan Center Frequencies:
3621.2, 7067.9, 7103.7, (P3), 10146.2 (P3), 14096.0 (P3)
VHF 1200 baud packet frequency for KF6NPC: 145.07

KN6KB - Rick in Rockledge, Florida, Scan Center Frequencies:
7068.9, 7103.7(P3), 10146.2(P3), 14066.4

KQ4ET – Joel, Virginia Beach, VA, Scan Center Frequencies:
3628.7, 7067.9, 10146.5(P3), 14110.0(P3)

K4CJX - Steve in Nashville, Tennessee, K4CJX Center Scan Frequencies:
Station # 1: 7076.9 (P2), 7101.2 (P3), 14076.9 (P2), 14106.7(P3)
Station # 2: 10123.9 (P2), 10141.2(P3)
Station # 3: 18103.9 (P2), 18108.7(P3)

K4SET - Scott in Murray, Kentucky, Scan Center Frequencies:
7074.9, 7103.7(P3), 10,136.9, 10143.4(P3), 21073.9, 21095.2 (P3)

K6CYC - Scott in Los Angeles, California, Scan Center Frequencies:
7069.9, 10123.9, 10143.7# - Omni-directional
21068.9, 21096.2(P3), 14068.9, 14102.7(P3) - Beaming South Pacific

W7IJ - Bill in Spanway, WA, Scan Center Frequencies:
Station 1: 3631.9(P3), 7068.9(P2), 7103.7 (P3), 10139.5 (P2)
Station 2: 14069.4(P2), 14110.0 (P3), 21077.9 (P2), 21091.2 (P2)

K6IXA - Grady in Atwater, California, Scan Center Frequencies:
Station # 1: 10122.9, 10143.7(P3)
Station # 2: 14064.9, 14102.7(P3)

K7AAE - Ronald in Woodinville, Washington, Scan Center Frequencies:
Station # 1: 3629.9, 7076.9, 10133.9, 10145.7(P3)
Station # 2: 14067.9, 14109.2(P3)

N8PGR, North Royalton, Ohio (20 miles south of downtown Cleveland and lake Erie)
Scan Center Frequencies: 3621.9, 7071.9, 10140.4, 14117.9

N0IA - bud in Deltona, Florida, Scan Center Frequencies:
3626.9, 7072.9, 10133.9, 14072.9, 14098.7(P3), 18106.2(P3)

WA2DXQ - Dave in Ft. Lauderdale, Florida

WB5KSD - Jon in Farmersville, Texas, Scan Center Frequencies:
7075.9, 10132.9, 14078.9, 14109.2 (P3)

WB0TAX - Deni in Elm Grove, Louisiana, Scan Center Frequencies:
7103.7 (P3), 10133.9, 10143.7 (P3), 14066.9 (P2), 14096.2 (P3), 18106.2 (P3)

WD8DHF - Gary in Harker Heights, Texas, Scan Center Frequencies:
Station #1: 3590.9, 7075.4, 70103.7(P3), 10,127.9
Station #2: 14075.4, 14098.7(P3), 18075.4, 18107.9(P3)
Station # 3: 21075.4, 21091.2 (P3)

WU3V - Jim in Great Falls, MT, Scan Center Frequencies:
3631.2 (ALL), 7074.9, 7103.7(P3), 10126.9, 10143.4(P3), 14069, 14102.7(P3)

WX4J - Earl in Switzerland, Florida, Scan Center Frequencies:
3622.4, 3620.9(P3), 7066.9, 7065.4(P3), 10143.4, 10141.9(P3) 14066.9, 14065.4(P3)

W1ON, Bedford, Massachusetts (near Boston), Scan Center Frequencies:
3620.9, 7070.9, 14075.9, 14104.2 (P3), 18100.9

W6IM - Rod in San Diego, California, Scan Center frequencies:
Dipole - 7073.9, 10141.2(P3)
Beam (135 deg) - 14073.9, 14098.7(P3)

W7BO - John in Woodland, Washington, Scan Center Frequencies:
7.067.9, 7101.2(P3)

W9GSS, East Peoria, Illinois, Scan Center Frequencies:
7072.9, 14073.9, 14109.9 (P3), 21098.0 (P3)

W9MR, Keensburg, Illinois, Scan Center Frequencies:
7065.9, 10145.2 (P3), 14069.9, 14101.7 (P3)

Currently Active Long Island VHF Packet Winlink 2000 Gateways

Log onto the nearest packet station. Then, connect to one of the gateways below. If a gateway is local and reachable, log on directly.

Hopping from node to node will slow the operation. From Northport, NY a connection needs to be made to the local WA2PNU digipeater, then to KC2COJ digipeater, then to WA2GUG-10 gateway. From Northport, use the direct link to N1EZT-10 gateway in Stamford, CT or N1BDF-10 gateway in New Haven, CT.

RMS Packet Gateway Positions Map is found at http://www.winlink.org/RMSPacketPositions

WA2GUG-10, Far Rockaway, NY, 145.090 MHz

K2YYD-10, Levittown, NY, 145.090 MHz

KC2UCD-10, Smithtown, NY, 145.070 MHz

KC2TGD-7, Montauk, NY, 145.070 MHz

N1EZT-10, Stamford, CT, 145.030 MHz

N1BDF-10, New Haven, CT, 145.050 MHz

W1GTT-10, New London, CT, 145.030 MHz

W1JPZ, Charles Town, CT, 145.050

Airmail Software Protocols

Software: Download and install "Airmail" version 3.1.936 or later.

Address Book Setup
    Main page
    Click on the black book icon (second from left)
    Click New
    Name: person's name
    To: Their email address
    Email gate: Email
    Post Via: WL2K
    Click OK

Formatting a New Message
    Main page
    Click on white page icon (third from left)
    Click on a name from your address book
    Click OK
    Type a message
    Click on the small floppy disk icon (6th from the left to save it)
    Click on the mailbox icon to post it (7th from the left to say it is ready to send)
    Now, click on the Outbox (on the left, you should see a mailbox with a blue arrow and the message) The message will be sent by whatever modality is used to make a connection.

Telnet Client Setup
    Internet connection to this computer is required
    Main page
    Click on Modules
    Click on Telnet Client
    Click on the yellow shaking hands
    Click on New
    Place in the boxes
    Remote Callsign: K4CJX
    Remote Host: k4cjx.no-ip.com
    Port: 12001
    Timeout: 30
    Local Callsign: your call
    Password: WL2KTELNETCLIENT
    Check B2
    Click OK
    You should see K4CJX in the window at the top of the Telnet Client now.
    Click the green button/icon and if you are connected to the Internet it should make a connection with Steve and sign off automatically.

Sending a Message Through Telnet
Main page
Click on modules
Click on Telnet Client
Click on the green button and the message should go. Watch closely, it does not take long.

VHF Packet Client Setup
Main page
Tools
Options
Modules
VHF Packet Client - check the block and click on the Setup button
From the Connections column:
TNC Type: select the TNC you are using - KPC-3 is at the bottom of the list
Com Port: Select your computers serial com port (usually Com1)
Baud Rate: Select the serial port baud rate (usually 9600)
Do not make changes in the Port Settings column
Check Show Hints
Check Terminal Window and Telnet Client
Check Show in Taskbar for Terminal Window and for VHF Packet Client and Telnet Client
Click Apply at the bottom
Click OK
Close Airmail and restart it.
Note: you cannot program multiple connections. Connect directly to a Telpac node.

Sending a Message Through VHF/UHF Packet
Main page
Click on modules
Click on Packet Client
Connect To: callsign of the Telpac Station
Connect As: your callsign
Click on the green button and the message should go. Watch closely, it does not take long.

Sending a Message Through VHF/UHF Packet via Multiple Nodes
Main page
Click on modules
Click on Packet Client
Connect To: callsign of the first packet station
Connect As: your callsign
Click on the Handshaking icon (third from the left)
Click on the green button
Once connected to the first packet station, Click on the Keyboard Icon (fourth icon from the left)
Using the connect command (c) in the lower window (C callsign) to connect to the next node

If you need the connect to more nodes, use the above steps until you come to the Telpac node

Once connected to the telpac node, click on the Handshaking icon (third from the left)

The message should go. It may take a long time depending upon the number of nodes used

HF Client Setup

Main page
Tools
Options
Connections
From the Modem (TNC) Connection column:
Modem type: select the modem you are using (PTC-IIex)
Com Port: select the com port you are using (usually Com1)
Baud Rate: Select the serial port baud rate (57600)
Center Frequency: 1500
Check USB
From the Modem (TNC) Connection column:
Select none if you do not have a remote control for your radio
Check Show Hints
Click Apply at the bottom
Click OK
Close Airmail and restart it.

Sending a Message Through HF

Main page
Click on modules
Click on HF Client
Connect To: callsign of the PMBO Station
Connect As: your callsign
Click on the green button and the message should go. Watch closely, it does not take long.

Note: the first you log onto an HF station, your callsign and license may need to be verified.

Satellite Radio

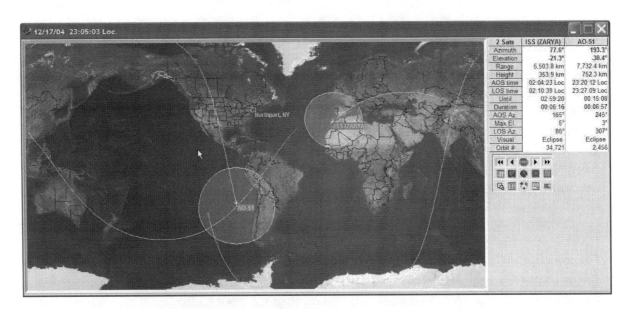

Amateur Satellite Radio (AMSAT) is basically a repeater or transponder in orbit around the earth. There are several satellites in Low Earth Orbit and several satellites in High Earth Orbit.

There are four basic categories of satellites:

1. Low Earth Orbit – Analog (CW and Voice)
2. Low Earth Orbit - Digital
3. High Earth Orbit
4. Occupied Spacecraft

Most of the amateur satellites and occupied spacecraft are in Low Earth Orbit (LEO). Low Earth Orbit satellites orbit the earth many times a day. Because Low Earth Orbit satellites have low orbits, and sensitive receivers, omni directional antennas can be used without substantial amounts of power. However, their passes are short and communications must consequently be short. Low Earth Orbit satellites typically have an approximate 90 to 100 minute period of evolution (time to make one orbit around the earth). Your communications window is approximately 8-20 minutes. Low Earth Orbit satellites typically have orbits of approximately 250 km to 1,000 km. The orbits of occupied spacecraft are typically below 500 km.

High Earth Orbit (HEO) satellites require beam antennas, azimuth/elevation rotators, computer tracking, and higher power radios. High Earth Orbit satellites will have longer passes and consequently longer communications capabilities. High Earth Orbit satellites typically have orbits of approximately 11 hours. High Earth Orbit satellites typically have orbits of 35,000 km at apogee and 4,000 km at perigee.

There are currently several analog satellites including RS-12, UO-14, RS-15, FO-20, AO-27, FO-29, SO-41, and the occupied spacecraft in orbit. The equipment required to communicate with these satellites are: an HF radio and/or a 2m / 70 cm radio. Packet equipment is required for digital work. Receiver preamps may also be needed.

Sending a transmission to a satellite is called an "uplink." Receiving a transmission from a satellite is called a "downlink." Uplink and downlink frequencies are different. Oscar is an acronym for "Orbiting Satellite Carrying Amateur Radio."

Working a satellite is very similar to working "split" on HF or "cross-band" repeat on repeaters, where you transmit on one band and listen on another. For example, if you chose RS-12, it will accept a signal anywhere from 145.910 MHz to 145.950 MHz and retransmit between 29.410 MHz and 29.450 MHz. These are known as the uplink and downlink passbands, and there is a direct relationship between them. A signal you transmit at 145.920 MHz will be retransmitted by the satellite at about 29.420 MHz and 145.930 MHz comes down as about 29.430 MHz, etc. This is because RS-12 (as well as RS-15) uses what is known as a "non-inverting linear transponder". The international space station uses uplink and downlink in the same (2 meter) band.

The following is a list of common satellite modes:

Dual Band Modes
A	2 meters uplink	10 meters downlink
B (UV)	70 cm uplink	2 meters downlink
J (VU)	2 meters uplink	70 cm downlink
K	15 meters uplink	10 meters downlink
LU	23 cm (1.2 GHz) uplink	70 cm downlink
LV	23 cm (1.2 GHZ) uplink	2 meters downlink
US	70 cm uplink	13 cm (2.4 GHz) downlink
LS	23 cm uplink	13 cm downlink
T	15 meters uplink	2 meters downlink

Single Band Modes
V	2 meters	2 meters
U	70 cm	70 cm
L	23 cm	23 cm
S	13 cm	13 cm
X	3 cm	3 cm

Uplink indicates the earth station transmit frequency. Downlink indicates the earth station received frequency.

Some satellites have dual modes that operate simultaneously. Satellites have 3 basic types of retransmissions: beacon, transponder, and repeater. Most satellites have a fixed Morse code beacon at the lower end of the satellites band-pass transponder. This is useful to detect when the satellite has crossed the horizon and is in range for operation. It can also be used to determine doppler shifts.

A transponder is similar to a repeater, but has a range of frequencies that are converted from one band to another. This range of frequencies is known as the pass band of the transponder. There are two types of transponders: Non-inverting and inverting. A non-inverting transponder will receive an upper side band signal at the high end of the uplink pass band and it will transmit it as an upper side band signal at the high end of the downlink pass band. An inverting transponder will receive an upper side band signal at the high end of the uplink pass band and it will transmit it as a lower side band signal at the lower end of the downlink pass band.

A repeater closely resembles an earthbound repeater. It listens for signals on one frequency and transmits on another frequency. All satellite repeaters (and transponders) are full duplex, meaning one can listen to the signal on the downlink while transmitting. Headphones are often used to avoid audio feedback.

Unlike earthbound communications where it is possible to pick a frequency and stay there, there is a phenomenon known as Doppler Shift that satellite that must considered. An example of Doppler shift is hearing a train blowing its whistle as it passed by? The tone changes as the train comes close and moves away. The sound inside the train remains the same. The change in tone is a result of the Doppler Shift. Signals coming from space experience the same phenomena as the satellite moves at a speed of about 17,000 miles per hour. The operator has to constantly tune the receiver and transmitter to make up the difference. The frequency shift varies by band. On RS-12 with its 2 meters uplink and 10 meters downlink, the change is about +/- 2.5 kHz. On FO-20 and FO-29, where the uplink is 2 meters and the downlink is 70 cm, the shift is about +/- 10 kHz.

Satellites travel in an elliptical orbit. Apogee is the point in a satellite's orbit where it is farthest from the earth. Perigee is the point in a satellite's orbit where it is closest to the earth. Inclination is the angle of the orbital plane with respect to the earth's equator. A node is the point where the orbital path crosses the equator. The ascending and descending pass is the south to north or north to south communications opportunity. These are also the points of Acquisition of Signal (AOS) and Loss of Signal (LOS). A footprint is the area of the earth's surface, which is visible to the satellite at one time. The lower the satellite's orbit, the smaller the footprint. Keplerian Elements (KEPS) are a set of numerical data that represents a satellite's orbital characteristics. The use of this information allows tracking programs to determine where that satellite is at any one time, to predict passes, and plot ground tracks. Keplerian elements should be updated every few weeks for stable orbits and more frequently if the object's orbit is altered. The AMSAT format is the most user-friendly format of Keplerian elements. However, the NORAD Two-Line Element (TLE) is the indigenous format available from NASA.

The AMSAT format looks like this:
    Satellite: AO-27
    Catalog number: 22825
    Epoch time: 03177.89093483
    Element set: 590
    Inclination: 98.2597 deg

RA of node: 200.9638 deg
Eccentricity: 0.0007455
Arg of perigee: 236.0777 deg
Mean anomaly: 123.9698 deg
Mean motion: 14.28984247 rev/day
Decay rate: 4.2e-07 rev/day^2
Epoch rev: 50820
Checksum: 344

Some communication satellites stay stationary with respect to the earth so that TV signals can be received with small antennas. To achieve this stationary position, a satellite must have a one-day orbital period like earth. The satellite to earth distance is calculated as 35,768 km. The satellite orbit must be circular with near zero inclination to stay stationary. If the orbit is closer to an inclined ellipse, the satellite projection to the earth surface will follow a skewed figure 8 pattern. This pattern is good for amateur radio because it can cover additional surface area of the earth and the satellite is closer to the earth at perigee.

Most of the time satellites are no higher than 35 degrees or so above the horizon. The closer the satellite is to the horizon, the greater the distance it is from the observer and the higher the path loss and the greater the transmit and receive gain that is needed. Vertical antennas will work well, especially ones with some gain, although some of the really high gain verticals are optimized for low angles of radiation and the signal strength falls off rapidly as the elevation angle increases. Another problem with verticals is that noise tends to be vertically polarized. This is not a problem with FM signals, but a big problem with SSB and CW. Dipole antennas also work well, but they often suffer a loss of gain off the ends. Beam antennas can be tilted up about 30 degrees to provide more gain toward the horizon. However, many satellite operators report excellent results in the standard flat horizontal orientation. When using beam antennas, the operator will need to continuously correct their direction as the satellite moves by. This becomes difficult to manage manually with the low earth orbit satellites because they have relatively fast velocities. Computer controlled antennas with azimuth/elevation rotators can track the satellites with ease. Azimuth/elevation rotators and computer controllers tend to be expensive.

A radio signal passing through the ionosphere changes polarization. A horizontally polarized signal transmitted from a satellite would change polarization when reaching earth. This phenomenon is called the Faraday Rotation. Circularly polarizing antennas are often used to deal with Faraday Rotation. Circularly polarized antennas will also minimize the spin modulation effect, which is caused by a satellites rotation of approximately 1 revolution per second.

Most satellites have an automatic transmitter at the satellite called the beacon. The beacon is usually located at the high or low end of the pass-band and will send out satellite identification and telemetry. Most beacons use CW.

When beginning satellite operation, try to get on one of the FM satellites, such as UO-14, AO-27, or ISS. These satellites are probably the easiest to work with a minimum amount of equipment. This can be done with an HT and a directional antenna. The antenna does not need to have much gain. It can be a 3-element beam that is held pointing at the satellite. Arrow antenna makes the 146/437-10 handheld antenna with a foam grip for about $75. Satellites like FO-20 and FO-29 are easy to work, but require two beams with an azimuth/elevation rotator and a computer controller. More sophisticated equipment will allow you to transmit and receive simultaneously on 2 bands and make frequency shifting for Doppler correction easier. At the time of this writing, AO-27 is only available on weekends. UO-14 is probably the best satellite to start with. ISS is easy to reach since it is so low. However, the ISS crew are not always available. Check the NASA website site for "Crew Scheduling" to see what times are allocated for amateur radio.

There are a number of tracking software programs on the market. After much trial and error by this author, the NOVA software program appears to be the best. It works well with Windows XP and the operator can update the Keplerian elements with the click of the mouse, provided there is with an internet connection. Some programs were actually inaccurate as compared to the online tracking by NASA of the international space station.

Online resources include:

> http://www.amsat.org
> http://www.amsat.org/amsat/news/wsr.html (weekly satellite status report)
> http://www.amsat.org/amsat/ftp/keps/current/amsat.all (Keplerian elements)
> http://www.arrl.org
> http://ariss.gsfc.nasa.gov
> http://spaceflight.nasa.gov/realdata/tracking/index.html (tracking international space station)
> http://www.nlsa.com/index.html (Nova satellite tracking software)
> http://www.orbitessera.com/ (N2WWD webpage)

Kilometer = Miles x 1.621388

Miles = Kilometers x 0.621388

Region 1: Africa, Europe, Russia, Middle East (excluding Iran), and Mongolia

Region 2: The Americas, including Hawaii, Johnston Island, and Midway Island

Region 3: The rest of Asia and Oceania

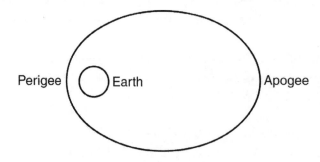

Low Earth Orbit – Analog Satellites (partial list)

Satellite	Frequencies (MHz)	Transponder /Beacon	Mode	Notes

RS-15 (NORAD 23439)
Launched: December 26, 1994 from the Baikonur Cosmodrome
Approximate height (varies) 2,160 km – 1,885 km

Downlinks	29.352	B		CW
	29.354 - 29.394	T	A	CW / USB
Uplinks	145.858 -145.898	T	A	CW / USB

Semi-operational as of June 8, 2003

FO-20 (Fuji-OSCAR 20, NORAD 20480) (2)
Launched: February 07, 1990 by an H1 launcher from the Tanegashima Space Center in Japan.
Approximate height (varies) 1,745 km – 912 km

Downlinks	435.795	B		CW
	435.800 - 435.900	T	J	CW / USB
Uplinks	145.900 - 146.000	T	J	CW / LSB

Operational as of June 8, 2003

AO-27 (OSCAR 27, AMRAD, NORAD 22825) (3)
Launched: September 26, 1993 by an Ariane launcher from Kourou,
   Approximate height (varies) 804 km – 790 km

Downlink	436.795		J	FM Voice
Uplink	145.850		J	FM Voice

Semi-operational as of June 8, 2003

FO-29 (Fuji-OSCAR 29, NORAD 24278) (also see 1200 and 9600 Baud)
Launched: August 17, 1996, by an H-2 launcher from the Tanegashima Space Center in Japan.
Approximate height (varies) 1,333 km – 801 km

Downlinks	435.795	B		CW
	435.800 - 435.900	T	J	CW / USB
Uplinks	145.900 - 146.000	T	J	CW / LSB
Downlink	435.910	B		FM Voice digitalker
Downlink	435.795	B		12 WPM CW telemetry

Operational as of June 8, 2003

SO-41 (SAUDISAT 1A, NORAD 26545)
Launched: September 26, 2000 aboard Soviet ballistic missile from the Baikonur Cosmodrome.
Approximate height (varies) 672 km – 614 km

Downlink	436.775		J	FM Voice
Uplink	145.850		J	FM Voice

Operational, but, intermittent as of June 8, 2003

Satellite	Frequencies (MHz)	Transponder /Beacon	Mode	Notes

SO-50 SAUDISAT-1C
Launched: December 20, 2002 aboard a Soviet ballistic missile from the Baikonur Cosmodrome.
Approximate height (varies) 650 km

Uplink	145.850 MHz (67.0 Hz PL tone)	T	J	FM Voice
Downlink	436.800 MHz	T	J	FM Voice

Operational as of June 8, 2003

AO-51 (ECHO)
Launched: July 30, 2004
Approximate height (varies) 718 km
Analog Uplink   145.920 MHz FM (PL - 67Hz), 1268.700 MHz FM (PL - 67Hz)
Analog Downlink   435.300 MHz FM, 2401.200 MHz FM
PSK-31 Uplink   28.140 MHz USB
Digital Uplink   145.860 MHz 9600 bps, AX.25, 1268.700 MHz 9600 bps AX.25
Digital Downlink 435.150 MHz 9600 bps, AX.25, 2401.200 MHz 38,400 bps, AX.25
Broadcast Callsign: PACB-11
BBS Callsign: PACB-12
Operational as of November 10, 2004

Low Earth Digital Satellites (partial list)

AO-16 (OSCAR 16, Pacsat, Microsat-A, NORAD 20439)
Approximate height (varies) 797 km – 780 km

Downlinks	437.025	T/B	J	1200 bps PSK SSB
	437.051	T	J	1200 bps PSK SSB
	2401.14280	B		1200 bps PSK SSB - (usually off)
Uplinks	145.900	T	J	1200 bps AFSK FM
	145.920	T	J	1200 bps AFSK FM
	145.940	T	J	1200 bps AFSK FM
	145.960	T	J	1200 bps AFSK FM

Semi-operational as of June 8, 2003

NO-44 (PC-SAT, NORAD 26931)
Approximate height (varies) 800 km – 790 km

Downlink	144.390		V	APRS
Uplink	145.827		V	1200 bps AX.25 AFSK

Operational as of June 8, 2003

Satellite	Frequencies (MHz)	Transponder/Beacon	Mode	Notes

FO-29 (Fuji-OSCAR 29, NORAD 24278) (see also Analog and 1200 Baud)
Approximate height (varies) 1,333 km – 801 km

Downlink	435.910	T	J	9600 bps FM BPSK
Uplink	145.850	T	J	9600 bps FM
	145.870	T	J	9600 bps FM
	145.910	T	J	9600 bps FM

Callsign 8J1JCS, Operational as of June 8, 2003

AO-51 (ECHO)
Approximate height (varies) 718 km
PSK-31 Uplink   28.140 MHz USB
Digital Uplink   145.860 MHz 9600 bps, AX.25, 1268.700 MHz 9600 bps AX.25
Digital Downlink   435.150 MHz 9600 bps, AX.25, 2401.200 MHz 38,400 bps, AX.25
Broadcast Callsign: PACB-11
BBS Callsign: PACB-12
Operational as of November 10, 2004

High Earth Orbit Satellites (partial list)

AO-10 (OSCAR 10, Phase 3B, NORAD 14129)
Launched: June 16, 1983 by an Ariane launcher from Kourou, French Guiana.
Approximate height (varies) 35,421 km – 4,026 km

Downlinks	145.810	B		CW
	145.825 - 145.975	T	B	CW / USB
Uplinks	435.030 - 435.180	T	B	CW / LSB

Semi-operational as of June 8, 2003

Occupied Spacecraft (partial list)

ISS (International Space Station)(NORAD 25544)(6)
ARISS (Amateur Radio on the International Space Station)(NORAD 25544)(6)
The ARISS initial station was launched September 2000 aboard shuttle Atlantis.
Approximate height (varies) 402 km – 391 km

Downlink	145.800	T	V	Worldwide voice/packet
Uplink	144.490	T	V	Region 2 & 3 voice
Uplink	145.200	T	V	Region 1 voice
Uplink	145.490	T	V	Worldwide packet TNC callsign: RS0ISS-1 U.S. callsign: NA1SS Russian callsigns: RS0ISS, RZ3DZR

Operational as of June 8, 2003

Antennas and Propagation

Any conducting material can act as an antenna. Designers go to "great lengths" to design antennas in order to control their radiation pattern and gain. The two main factors in antenna design and operation are the geometry of the antenna and the proximity of the antenna to nearby objects.

The Half Wave Dipole

Each half of the half wave dipole is 1/4 wavelength. Together they make up 1/2 wavelength.

The free space wavelength of an electromagnetic wave is: L = c / f (Hz), where c = 300,000,000 meters/second (velocity of light) or L (meters) = 300 / f (MHz).

The velocity of a wave along an antenna or transmission is slower than it is in free space, usually about 95 % of c. Therefore, a half wave dipole is 5 % shorter than it's free space wavelength.

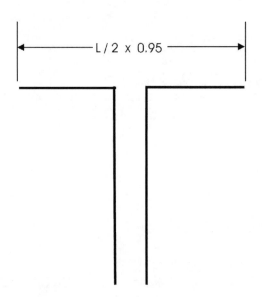

A 7 MHz. half wave dipole length = (c / f) x 0.95 x 0.5 meters

A 7 MHz. half wave dipole length = (300 / 7) x 0.95 x 0.5 = 20.36 meters

A 7 MHz. half wave dipole length = ((c / f) x 0.95 x 0.5) x 3.2808 feet

A 7 MHz. half wave dipole length = ((300 / 7) x 0.95 x 0.5) x 3.2808 = 66.79 feet

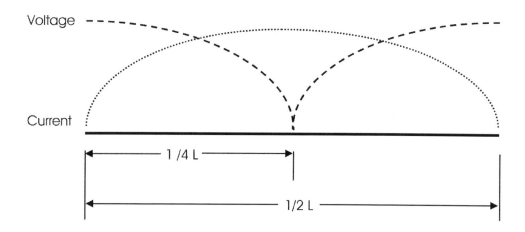

The voltage and current waves of an antenna are 90 degrees out of phase with each other.

If the current is high and the voltage is low, the impedance will be low (R = E/I). The impedance is the lowest at the center of a half wave dipole (72 ohms). At the ends, the impedance is the highest (2,000-3,000 ohms).

The half wave dipole is resonant when the length is made such that the mid point yields the lowest voltage and highest current.

A 1/4 wavelength antenna or an antenna of multiple 1/4 wavelengths is a resonant antenna.

Any circuit is resonant when the inductive reactance and capacitive reactance are equal. At resonance, inductive reactance and capacitive reactance are 180 degrees apart and when equal, cancel each other, leaving only the resistance component. The shortest length of wire that can be resonant is a quarter wavelength.

If the antenna is shorter than a quarter wavelength, it will have a capacitive reactance. It will require the addition of an inductive reactance (loading coil) to cancel the capacitive reactance and become resonant.

If the antenna is longer than a quarter wavelength and shorter than a half wavelength, it will have an inductive reactance. It will require the addition of a capacitive reactance (capacitor) to cancel the inductive reactance and become resonant.

The Five-Eighth Wave Ground Plane Antenna

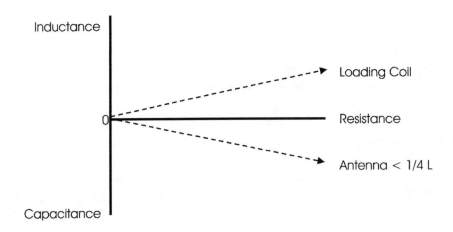

Normally, a 5/8 L antenna cannot be resonant because it is not a multiple of a 1/4 L. However they are commonly found as vertical antennas. The vertical section is 5/8 L, which has a capacitive reactance because it is longer than two 1/4 L's. An inductive loading coil is added somewhere along the length of the vertical to cancel the capacitive reactance. The 5/8 wave antenna is actually tuned to 3/4 L by the addition of the loading coil making it effectively a 3/4 wavelength antenna. The four horizontal radials are 1/4 wavelength each.

The Quarter Wave Ground Plane Antenna

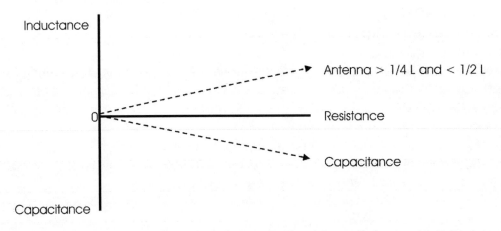

The quarter wave ground plane antenna is often used with radials or installed on a flat conducting surface, such as the roof of a car. It is simply a reflector whereby some of the radiation will be reflected by the ground plan and interact with the incident wave from the antenna. This kind of antenna provides an omni directional pattern.

The Yagi Antenna

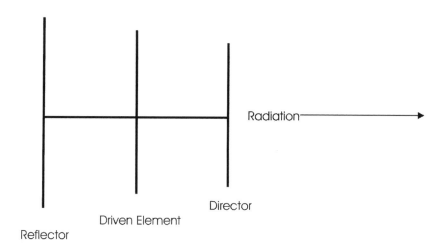

A Yagi antenna has one driven element, a dipole, and at least one other parasitic element. The parasitic dipoles receive radiation from the main dipole and re-radiate it. The energy is radiated from the main driven element and after a short delay, energy is picked up and re-radiated by the parasitic elements. The antenna can be made to radiate in one direction by controlling the spacing and length of the parasitic elements to yield a greater gain.

The three element Yagi has: a driven element, a reflector element, and a director element. The primary direction of radiation is in the direction of the driven element to director element. The driven element length is determined by equations used to calculate the dipole antenna. The reflector is 5% longer than the driven element and the director is 5% shorter than the driven element. The spacing between the elements are usually between 0.15 to 0.2 wavelengths (0.18 for maximum forward gain).

Propagation delays caused by the spacing of the elements, causes wave cancellation towards the rear of the antenna and wave reinforcement towards the front of the antenna. The gain is usually about 6-8.5 dB over a dipole.

Additional elements can be added to a Yagi to increase the power gain. Additional elements are always directors placed in front of the driven element. Doubling the number of directors, will increase the gain by about 3 dB. Adding parasitic elements to a Yagi decreases the antenna bandwidth.

The Folded Dipole Antenna

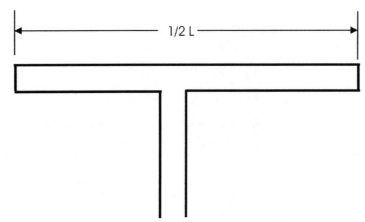

The folded dipole antenna is an antenna that consists of two dipoles connected in parallel. Folding the dipole increases the feed point impedance. Two dipoles folded, the feed point resistance increases by 2 squared (4), which is equal to 300 ohms. Three parallel dipoles increases by 2 cubed (8). The dipole forming the driven element of a Yagi antenna is often folded to increase the feed point impedance by 4.

Trapped Antennas

Traps are parallel tuned circuits used on HF antennas that have high impedances used to switch different sections of the antenna in and out of resonance. When the traps are not used at their resonant frequencies, they become loading coils and shorten the effective length of the antenna wire needed.

Isotropic and Practical Antennas

The isotropic antenna is a theoretical antenna that radiates equally in all directions and serves as a reference. A half wave dipole antenna has a gain of 2.14 dB above an isotropic antenna. Rubber duck antennas have a loss of several dB below an isotropic antenna.

An antenna that directs most of the electromagnetic radiation in one direction has a power advantage over an omni directional antenna. The power gain of an antenna is the ratio of the power radiated in its primary direction as compared to the isotropic antenna. A gain of 6 dB is a factor of 4 and 1 "S" point on the S-meter.

The area of a dipole's radiation pattern (A1) is 1.64 times that of the isotropic's pattern (A2). Therefore, the gain can be calculated as Gain = 10 Log (A1/A2) = 10 Log (1.64/1) = 2.15 dB

See radiation patterns on the next page.

Voltage Standing Wave Ratio (VSWR)

The Voltage Standing Wave Ratio (VSWR) is the ratio between the impedances of the feed line and the load. If we connect a 50 Ohm resistor at one end of a piece of 50 Ohm coaxial cable, and connect a transmitter and SWR meter at the other end, the VSWR will be 1:1. The resistor is not resonant. However, if we connect a resonant antenna that has an impedance of 144 Ohms to the end of that piece of cable, the VSWR will be 2.88:1 (VSWR = Antenna Impedance / Feed Line Impedance).

If a feed line is cut to a length that creates a VSWR measurement of 1:1 at the transmitter end of that feed line, the actual VSWR on this line is (infinity):1. Using VSWR is not the best method for tuning an antenna. The best method to measure the resonant frequency of an antenna is to use an antenna bridge at the antenna.

High VSWR does not cause feed line radiation. Most of the radiation from a coaxial cable is caused by terminating an unbalanced feed line with a balanced load. The remainder of the radiation is due to other problems such as, braid corrosion, improperly installed connectors, and signal pickup caused by routing the feed line too close to, and parallel to the antenna.

A properly terminated and installed open wire line does not radiate. Even with infinite SWR, the fields surrounding each wire cancel each other out. Terminating the line in an unbalanced load, or causing anything to come within the "field space" will cause unbalance in the line, thus allowing the line to radiate.

Propagation

The ionosphere is a layer in the Earth's atmosphere that lies in a range of 80 to 300 miles above the Earth's surface that reflects radio waves. As the sun shines on the ionosphere it changes composition and height, which affects the propagation characteristics. In general signals below 30 MHz bounce off this layer and return to Earth while signals above 30 MHz go through the layer into outer space. Radio signals that are bounced or refracted off the ionosphere are also affected by the time of day and season of the year.

During the 24-hour cycle the ionosphere changes in height above the Earth and bounces some signals while absorbing others. During the day the higher frequencies (above 10 MHz) tend to propagate while lower frequencies are absorbed. At night the reverse happens. There are many exceptions to this but it is a good general guideline.

Seasons also affect propagation. Summertime in the northern hemisphere means that higher frequencies have better propagation while in the winter the lower frequencies improve. An interesting time of the year for propagation is when the seasons change from fall to winter and from winter to spring. This is often when the best DX can be found. Because the seasonal change is occurring in both hemispheres but in the opposite direction DX from North American to Australia or southern Africa can be at its best.

Another phenomenon that affects radio propagation is the 11-year sunspot cycle. A peak occurred during the year 2000 and the next peak will occur around 2011. A sunspot low occurs at the midpoint of this cycle. When the sunspots are at their maximum propagation is at its best. At this time the higher shortwave frequencies exhibit the best propagation extending to 6 meters, which becomes quite popular during this time of the cycle. 10 meters can easily work stations worldwide with low power (even qrp) and a modest antenna.

Band (meters)	Frequency (MHz)	Use (band conditions vary for many reasons)
160	1.8 – 2.0	Night
80	3.5 – 4.0	Night and Local Day
40	7.0 – 7.3	Night and Local Day
30	10.1 – 10.15	CW and Digital
20	14.0 – 14.350	World-wide Day and Night
17	18.068 – 18.168	World-wide Day and Night
15	21.0 – 21.450	Primarily Daytime
12	24.890 24.990	Daytime During Sunspot Highs
10	28.0 – 29.7	Daytime During Sunspot Highs
6	50 – 54	Local to World-wide
2	144 – 148	Local and Medium Distance
70 cm	430 – 440	Local

Near Vertical Incidence Sky wave (NVIS) Antennas

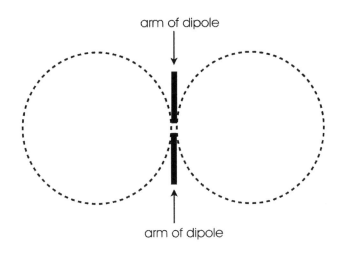

Dipole Radiation Pattern

NVIS propagation is a propagation pattern that uses antennas with high-angle radiation (almost 90 degrees, vertical) and low operating frequencies for a range of about 300 miles.

Long distance propagation uses radio waves that are reflected from the ionosphere and return to earth at some distance away. Radio waves that are radiated at a very low angle, travel a long distance to reach the ionosphere at a very shallow angle and return to earth far away. When the angle of radiation increases, the radio waves reach the ionosphere at a greater angle, and return to earth closer to their point of origin. Signals that reach the ionosphere at a higher angle of incidence will not be reflected at all, but will continue out into space. The area of reflection that would have occurred is the "skip zone". Depending on operating frequencies, antennas, and propagation conditions, this skip zone can start at roughly 12 to 18 miles and extend out to several hundred miles, preventing communications.

NVIS antennas are designed to minimize the ground wave (low takeoff angle) radiation and maximize the sky wave (very high takeoff angle, 60-90 degrees). Essentially, the NVIS antenna radiates a wave almost straight up, then bounces from the ionosphere and returns to the Earth in a circular pattern around the transmitter. Because of the near-vertical radiation angle, there is no skip zone. Communications are continuous out to several hundred miles from the transmitter. The nearly vertical angle of radiation requires the use of lower frequencies, usually 2-10 MHz. This type of propagation is excellent when communicating over hills and mountains. These frequencies are the same frequencies that contain a lot of atmospheric noise, such as distant thunderstorms. The NVIS antenna is optimized for listening to signals from nearby areas, and minimizes the reception of signals from distant sources.

One of the most effective antennas for NVIS is a dipole that is mounted from 0.1 to 0.25 wavelengths above ground. When a dipole is brought very close two ground, the angle of radiation increases. In the range of 0.1 to 0.25 wavelengths above ground, vertical and nearly vertical radiation reaches a maximum. A dipole can be used at even lower heights, resulting in some loss of vertical gain, but often, a more substantial reduction in noise and interference from distant regions. Heights of 5 to 10 feet above ground are not unusual for NVIS operation.

During a test by W0IPL, they used a 75-meter dipole at a height of 30 feet. They found the communications to be difficult. They set up a second dipole at a height of 8 feet. The background noise went from S7 to S3 and the communications with stations 25 miles and further, greatly improved. Many people find the 10 to 15 foot height to be ideal. Field tests have proven that the maximum NVIS efficiency is obtained at the 10 to 15 foot height for frequencies in the 40 meter to 75 meter range.

Skip Zones

A skip zone is the area that is not covered by sky wave radiation. In other words, the sky wave angle is such that the sky wave travels a long distance before reaching earth. The distance between the transmitting antenna and the point where the sky wave reaches the earth is the skip zone.

The "take-off angle" is the angle at which a wave leaves the transmitting antenna. Nomograph for determining primary skip zone (one hop) as a function of radiation angle.

For a 2 or 3 element yagi at HF (7 to 14 Mhz) approx heights are:
(assuming average soil conductivity = 5 mS / meter   dielectric constant = 13)

TO Angle (degress)	Height (feet)	Distance F2 Layer (miles)
45	20	400-800
40	27	450-900
30	32	650-1,300
20	37	950-1,700
15	45	1,200-2,000
12	50	1,300-2,300
10	60	1,400-2,400

For a 2 or 3 element yagi at HF (21 to 28 Mhz) approx heights are:
(assuming average soil conductivity = 5 mS / meter   dielectric constant = 13)

TO Angle (degress)	Height (feet)	Distance F2 Layer (miles)
45	20	200-400
40	27	225-450
30	32	325-650
20	37	475-850
15	45	600-1,000
12	50	675-1,150
10	60	700-1,200

H.F. Antenna Analysis

Theoretical Antenna Patterns

1/2 Wave Horizontal Dipole Antenna - 7 MHz

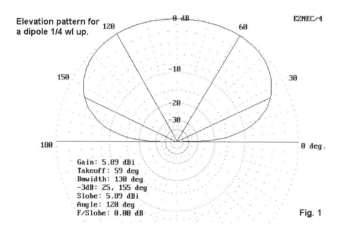

Fig. 1

Height (feet)	Take Off Angle (deg)
17.2	90
34.4	61
51.6	38
68.8	28
86.0	22
103.2	19
120.4	16
137.6	14

Maximum gain (5.89 db) is obtained at 59 degrees.

Yagi Horizontal Antenna – 7 MHz

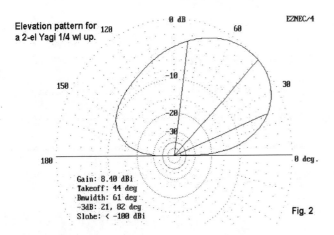

Fig. 2

Height (feet)	Take Off Angle (deg)
	58
34.4	45
51.6	34
68.8	27
86.0	22
103.2	18
120.4	16
137.6	14

Maximum gain (8.40 db) is obtained at 44 degrees.

The Yagi improvement in elevation angle is mostly at the lowest heights. Above a half-wavelength, the take-off angle closely matches that of the dipole. The Yagi exhibits about 3 dB more gain than the dipole at the angle of maximum radiation, which is 44 degrees.

The Vertical Dipole Antenna – 7 MHz
10' off the ground at lowest point
Feed point = 44'
Top = 78'

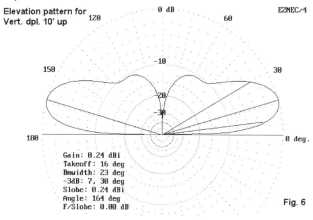

Fig. 6

Maximum gain (0.24 db) is obtained at 16 degrees.

1/4 Wave Vertical Antenna Counterpoise

Band (meters)	Length (meters)	Length (feet)
10	2.5	8.2
12	3	9.84
15	3.75	12.3
17	4.25	13.94
20	5	16.4
30	7.5	24.6
40	10	32.8
60	15	49.2
80	20	65.6
160	40	131.2

1/2 Wave Vertical Antenna Counterpoise

Band (meters)	Length (meters)	Length (feet)
10	5	16.4
12	6	19.68
15	7	22.96
17	8.5	27.88
20	10	32.8
30	15	49.2
40	20	65.6
60	30	98.4
80	40	131.2
160	80	262.4

Feet = 3.28 meters

Comparison of 1/2 Wave Horizontal Dipole and Yagi Antennas - 7 MHz

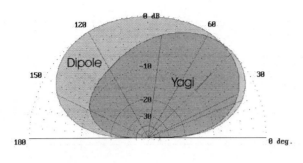

1/2 Wave Horizontal Dipole and Yagi

Fig. 9

- Horizontal dipole
- Horizontal Yagi

The horizontal dipole's maximum gain (5.89 db) is obtained at 59 degrees.

The horizontal Yagi's maximum gain (8.40 db) is obtained at 44 degrees.

The Yagi has slightly more gain in the lower angles of radiation. The horizontal dipole has a wider range.

Comparison of 1/2 Wave Horizontal Dipole, Yagi, and Vertical Antennas - 7 MHz

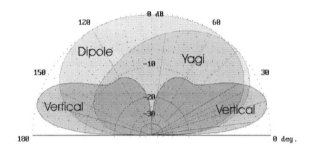

1/2 Wave Horizontal Dipole, Yagi, Vertical

Fig. 10

- Horizontal dipole
- Horizontal Yagi
- Vertical Dipole

The horizontal dipole's maximum gain (5.89 db) is obtained at 59 degrees.

The horizontal Yagi's maximum gain (8.40 db) is obtained at 44 degrees.

The vertical's maximum gain (0.24 db) is obtained at 16 degrees.

The horizontal dipole has the widest overall coverage.

The Yagi offers slightly more gain at slightly lower radiation angle than the horizontal dipole.

The vertical has more gain at lower radiation angles, excellent for long-range transmission.

Skip Distance, Skip Zones, and "Take Off" Angles of Radiation

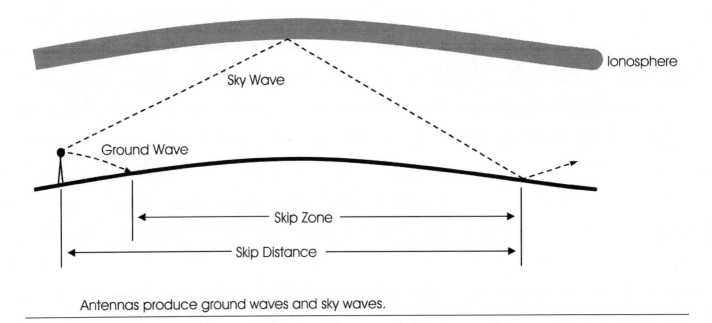

Antennas produce ground waves and sky waves.

A sky wave is a signal that travels toward the ionosphere and is reflected back down to earth. HF sky waves typically travel 100 to 8,000 miles. VHF sky waves typically travel 50 to 150 miles. The angle at which the sky wave is sent from the antenna to the ionosphere is called the "take off angle of radiation." The angle of radiation or take off angle is dependent upon the antenna type, the height of the antenna, and the frequency of the electromagnetic wave.

A ground wave is a signal that runs along the Earth's surface. It extends out from the antenna for up to about 50 miles. It is a limited signal, which allows for short distance communication.

A skip zone is an area where no signals will be received. Skip zones are formed when the nearest point at which a sky wave is received is beyond the furthest point at which a ground wave is received. When the ground wave coverage is great enough or the skip distance is short enough that no zone of silence occurs, there is no skip zone.

The ionosphere will reflect frequencies from 0.1 to 30 MHz.

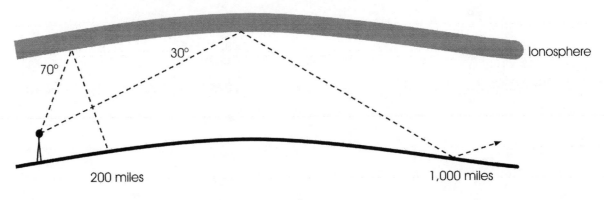

Empirical Data

Nassau Amateur Radio Club
Skip Zone (one hop) as a Function of Radiation Angle – 7 - 14 MHz. 2-3 element Yagi

Height (feet)	Take Off Angle (deg)	Skip Zone F2 Layer (miles)	Miles Skipped
20	45	400-800	400
*27	40	450-900	450
*29.5	35	675-975	300
*32	30	650-1,300	650
37	20	950-1,700	750
45	15	1,200-2,000	800
50	12	1,300-2,300	1,000
60	10	1,400-2,400	1,000

* Best suited for Field Day. Skip distance is about 1,500 miles, which allows optimum coverage of the East coast where most of the Field Day contacts are. It would also still allow one hop coverage to the West coast (about 6 to 10 db down for a 2 or 3 element Yagi), with a double hop zone of about 12 to 18 db down due to multi-skip losses over mid US terrain.

Nassau Amateur Radio Club
Skip Zone (one hop) as a Function of Radiation Angle – 21 - 28 MHz. 2-3 element Yagi

Height (feet)	Take Off Angle (deg)	Skip Zone F2 Layer (miles)	Miles Skipped
10	45	400-800	400
*13.5	40	450-900	450
*14.8	35	675-975	300
*16	30	650-1,300	650
18.5	20	950-1,700	750
22.5	15	1,200-2,000	800
25	12	1,300-2,300	1,000
30	10	1,400-2,400	1,000

* Best suited for Field Day. Skip distance is about 1,500 miles, which allows optimum coverage of the East coast where most of the Field Day contacts are. It would also still allow one hop coverage to the West coast (about 6 to 10 db down for a 2 or 3 element Yagi), with a double hop zone of about 12 to 18 db down due to multi-skip losses over mid US terrain.

Selecting an H.F. Antenna for Field Day

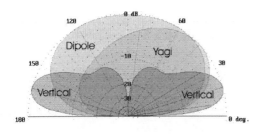

1/2 Wave Horizontal Dipole, Yagi, Vertical

Fig. 10

As indicated in Figure 10 above, the 1/2 wave vertical antenna will produce very good long-range communications because of its low angle of radiation. The 1/2 wave horizontal Yagi antenna will produce good long-range communications because it has slightly more gain in one direction and it has a lower angle of radiation. It also has much less gain in the reverse direction. The 1/2 wave horizontal dipole antenna will produce the best overall communications near and far because it has a wide angle of radiation in the forward and rear directions.

The horizontal 1/2 wave dipole will be chosen for use at Field Day because it has the best overall communications near and far. Most clubs are currently using dipole antennas for Field Day for the obvious reasons. As a secondary antenna, the Outbacker Outreach, which has been used in past Field Days, will continue to be used since it only takes a few minutes to set up. The Outbacker antenna is used by FEMA, U.S. Coast Guard, other military organizations, embassies and more.

There are two basic types of dipole antennas. The simple wire type that can be strung between two trees or a rigid aluminum tubing rotatable 1/2 wave dipole such as the Cushcraft D3 (10, 15, 20 meter) tri-band antenna. The rotatable dipole will facilitate communications in any direction by turning the mast. Either of these dipoles will provide excellent communications. . The Van Gorden wire dipoles have factory-installed balums. The Van Gorden D-20 (20 meter) wire dipole measures 33 feet in length.

Masts are available in many sizes and shapes. The Cushcraft D3 antenna weights only 9 pounds. Therefore a tower is not necessary. The mast can be rotated manually. Therefore, a 115 VAC powered rotator is not required. Furthermore, if there was an emergency, 12 volts may be the only power available. A mast of 24 feet can be constructed from 1-1/4" thick-wall (0.065") aluminum tubing and a public address speaker stand. A special guy ring can be made to allow the mast to be turned. There are also a number of different masts that can be obtained commercially or as military surplus.

Emergency Power

Emergency Power Requirement Estimate

Appliance	Rated Watts	x Hours/day	Surge Watts	Surge Watts
Light Bulbs (75 watt each)	75 x number	_____	75 x number	_____
Compact fluorescent (25/100 watt)	25 x number	_____	25 x number	_____
Refrigerator/Freezer	500	_____	2000	_____
Sump Pump	800	_____	2000	_____
Water Pump (1/3 HP)	1000	_____	3000	_____
Furnace Fan (1/2 HP)	875	_____	2300	_____
Electric Blanket	400	_____	400	_____
Space Heater	1800	_____	1800	_____
Heat Pump	4700	_____	12000	_____
Dehumidifier	650	_____	800	_____
Attic Fan	300	_____	900	_____
Table Fan	800	_____	2000	_____
Window Air Conditioner	1200	_____	4800	_____
Central Air (10k BTU)	1500	_____	6000	_____
Central Air (24k BTU)	3800	_____	15000	_____
Central Air (40k BTU)	6000	_____	24000	_____
Computer	300	_____	300	_____
CD Player	100	_____	100	_____
VCR	100	_____	100	_____
Radio	100	_____	100	_____
Television	300	_____	300	_____
Receiver	420	_____	420	_____

Microwave	800	_____	800	_____
Blender	300	_____	900	_____
Coffee Maker	1500	_____	1500	_____
Electric Range (1 element)	1500	_____	1500	_____
Toaster (2-slice)	1000	_____	1600	_____
Dishwasher (Hot Dry)	1500	_____	3000	_____
Electric Oven	3410	_____	3410	_____
Steam Iron	1200	_____	1200	_____
Washing Machine	1150	_____	3400	_____
Gas Clothes Dryer	700	_____	2500	_____
Electric Clothes Dryer	5400	_____	6750	_____
Security System	500	_____	500	_____
Deep Freezer	500	_____	1000	_____
Hair Dryer	1200	_____	1200	_____
Garage Door Opener (1/3 HP)	750	_____	750	_____
Electric Water Heater	4000	_____	4000	_____

Total Per Day   _____

_____

Note: the generator must be able to handle the total surge power.
Note: Compact fluorescent (25/100 watt) bulbs provide an equivalent of 100 watts of light and use 25 watts of power.

Generators

Generators are basically gasoline, natural gas, or propane powered. They usually generate substantial amounts of power. Portable generators commonly generate 1,000 to 5,000 watts continuously with a surge of about 1,300 to 6,500 watts. There are several disadvantages of gasoline powered generators. They require a constant refilling of gasoline and gasoline cannot be stored for long periods of time. Gasoline stations require electric pumps to supply gasoline and they may not have emergency generators. Natural gas is often available in many homes. Propane can be stored. Propane tanks are usually refilled by gas pressure, which eliminates the need for electric pumps.

Solar Power

Solar power is far more expensive than generator power. There are several advantages of using solar power. Sunshine is the source of power, which eliminates dependency on vendors for fuel. Solar power is clean and requires little maintenance. The disadvantage is initial cost. Typically, solar power is used to charge batteries, which are connected to an inverter.

A solar cell or photovoltaic cell (PV) is made of semiconductors, usually silicon. Ordinarily pure silicon is a poor conductor of electricity so impurities such as phosphorus and boron are added to create the semi-conductor. The addition of these impurities allows the silicon to conduct electricity. The semiconductor absorbs part of the light. The absorbed light energy knocks electrons loose, allowing them to flow freely. Metal contacts are placed on the top and bottom of the solar cell so that current can be drawn from it.

A solar electric panel consists of an aluminum framed sheet of highly durable low reflective, tempered glass that has had individual solar cells adhered to the inner glass surface. These individual solar cells are wired together in a series parallel configuration to obtain the necessary voltage and current. The back of the panel is protected by another sheet of tempered glass or a long lasting material such as Tedlar. The series parallel connections are passed through the protective backing and then wired to a weather proof junction box which is permanently mounted to the back of the panel where the panel's output connections are made. There are also flexible cells and panels, roof tile cells, etc.

Solar panels are rated as watts per hour. For example, in direct sunlight, a 50 watt solar panel will produce 50 watts per hour. It will produce 350 watts in 7 hours, and so on.

Batteries

When designing a marine deep cycle battery, manufacturers must keep in mind that the battery may be used for starting a boats engine. In order to start an engine, the battery must contain a lot of plates and plate area, which give the battery its high cranking capacity. In order to squeeze enough plates into a standard battery case, the plates must be made thin. The thinner the plates the shorter the life span of the battery when it is used in a deep cycle application. If cost is a major factor and the batteries will only be used occasionally during an emergency, a marine deep cycle battery may be adequate.

A much better choice for long-term continuous use is the golf cart battery. The plates are much thicker and designed to be deep cycled below 50% depth of discharge day in and day out, year after year. A properly maintained golf cart battery should last 3 two 5 years in a typical renewable energy application. A typical golf cart battery is available in a 6 volt 220 amp hour ratings. Two batteries will be required and they will need to be wired in series to produce 12 volts @ 220 amp hours. Golf cart batteries are considered the minimum type of battery that is used in renewable energy application. There are larger batteries available in 6, 4 and even 2 volt configurations which have even larger plates and thus longer life expectancies.

When there is power available from the utility company, batteries can be charged from the power line. During emergencies, when there is no power from the utility company, batteries will have to be charged from the solar panel.

Calculating Battery Usage:

$$\frac{\text{Amp Hour} \times \text{Volts (Watts)}}{\text{Hours of Use}} = \text{Watts per Hour}$$

$$\frac{\text{Amp Hour} \times \text{Volts (Watts)}}{\text{Watts per Hour}} = \text{Hours of Use}$$

Example:
A 12 volt, 100 ah battery will provide:
100 amps in 1 hour (1200 watts in 1 hour)
14.3 amps for 7 hours (171 watts in 7 hours)
5 amps for 20 hours (60 watts in 20 hours)

Calculating the Load:

Watt Hours = Load Watts x Hours of Use
Add 10 percent for battery losses.

Example: If a television draws 200 watts and runs for three hours (200 x 3 = 600) it will use 600 watt hours

Inverters

An inverter is an electronic device, which inverts DC energy AC energy. Most household appliances such as refrigerators, TVs, lighting, stereos, computer etc., all run off of AC electricity.

Modern DC to AC inverters are very reliable, quiet, and require virtually no maintenance. There are two different types of DC to AC inverters in common use today. The first type of inverter is known as a modified sine wave inverter. This type of inverter is very high in efficiency and produces a waveform, which is an approximation of the pure sine wave waveform.

High frequency units take the incoming 12 Volts DC and will step up that voltage to approximately 200 volts DC through a high frequency DC to DC converter circuit and then will take the 200 Volts and will wave shape it into a modified sine wave using a using a device called a high voltage H-bridge. The high voltage H-bridge is basically a group of field effect transistors that are arranged in such a way as to form the necessary half cycles that create the modified sine wave at the 60 Hz frequency required for US appliances. By utilizing high frequency, the need for a large iron core output transformer is eliminated and much smaller transformers can be used. As a result of this, high frequency inverters tend to be much lighter but do have a lower surge capacity because they lack the fly wheel effect found in heavy iron core output transformer based inverters.

Low frequency units take the incoming 12 Volts DC and converts it into AC, using a multivibrator or microprocessor based circuit. The AC is kept at a low voltage and is converted into a 60 Hz signal before it is fed to the iron core transformer. Wave shaping and the increased current that is needed to drive the transformer is performed again by an H-bridge which is a group of field effect transistors that are arranged in such a way as to feed high current pulses to the primary windings of the transformer at precise moments of each wave form half cycle. The transformer converts the lower voltage which was fed to its primary windings into 120 Volts AC at its secondary windings using simple transformer step up principles involving a 10 to 1 ratio, converting 12 Volts AC to 120 AC. This type of inverter is more durable than the high frequency inverters, and has a much higher surge capacity. Low frequency units tend to cost two to five times more than do high frequency units and often weigh four times more.

The second type of inverter is known as a pure sine wave inverter. This type of inverter produces pure sine waves, but at the cost of some efficiency loss and at a much higher price. Most pure sine wave inverters are typically priced at least 75% higher than the modified sine wave counterparts and in some cases do not have as high of a surge capability as do modified sine wave units.

A 3,000 Watt, 120 VAC, Output Solar System

The Batteries will need to supply 3,600 watts of electricity per hour, 86,400 watts per day. The inverter efficiency is about 71 percent. Therefore 3,600 watts DC will be needed to convert to 3,000 watts AC. The solar panels will need to charge the batteries with 86,400 watts during sunlight (about 7 hours per day) at the rate of 12,342.9 watts per hour. The system will require a space of about 30 feet x 30 feet, weigh more than 7,000 pounds, and cost over $60,000.

Calculate Time of Battery Use:

$$\frac{\text{Amp Hour} \times \text{Volts (Watts)}}{\text{Watts per Hour}} = \text{Hours of Use} = \frac{400 \text{ ah} \times 6 \text{ volts} \times 36 \text{ batteries}}{3600 \text{ watts per hour } (150 \text{ a} \times 24 \text{ v})} = 24 \text{ hours}$$

Calculate Time of Battery Recharge from Solar Panels:

$$\frac{\text{Amp Hours} \times \text{Volts (Watts)}}{\text{Solar Panel Watts}} = \frac{400 \text{ ah} \times 6 \text{ volts} \times 36 \text{ batteries}}{66 \times 185 \text{ watts per hour}} = 7.1 \text{ hours}$$

The system consists of:
- Sixty Six 185 watt, 24 v solar panels, 62" x 32.5" x 1.8", 38 pounds, (about $50,000) Connected in parallel (923.6 square feet [30' x 30'], 2,508 pounds)
- 3,600 watt output, 24 volt - 210 amp input (150 amp input at 3000 watt output), modified sine wave inverter (about $1,800)
- Three 60 amp charge controllers (about $600)
- Thirty six 400 ah, 6 volt, 127 pound batteries connected in series and parallel for 24 v (about $7,700 and 4,572 pounds)
- Miscellaneous cables, etc.

Anderson Powerpoles[R]

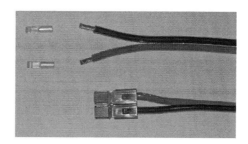

Figure 1

Figure 2
Connectors should be aligned in this direction
(Red on the Right with the conductor on the top)

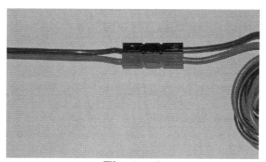

Figure 3

One of the nicest features of Anderson Powerpoles[R] is that there are no male or female connectors. To connect them together, simply turn one or one set of connectors upside down. This makes them universal and can be used to connect various pieces of equipment easily, even to someone else's power supply.

Strip off 1/4" of insulation and slightly tin the wires. Solder the metal connectors to the wires. I find crimping makes it very difficult if not impossible to insert the metal pieces into the plastic shell. After soldering the wires to the metal connectors, allow them to cool. Then insert them into the plastic shells as seen in Figure 2.

To connect to sets of cable together, simply turn one pair upside down and push together as seen in Figure 3.

D-Star and Programming

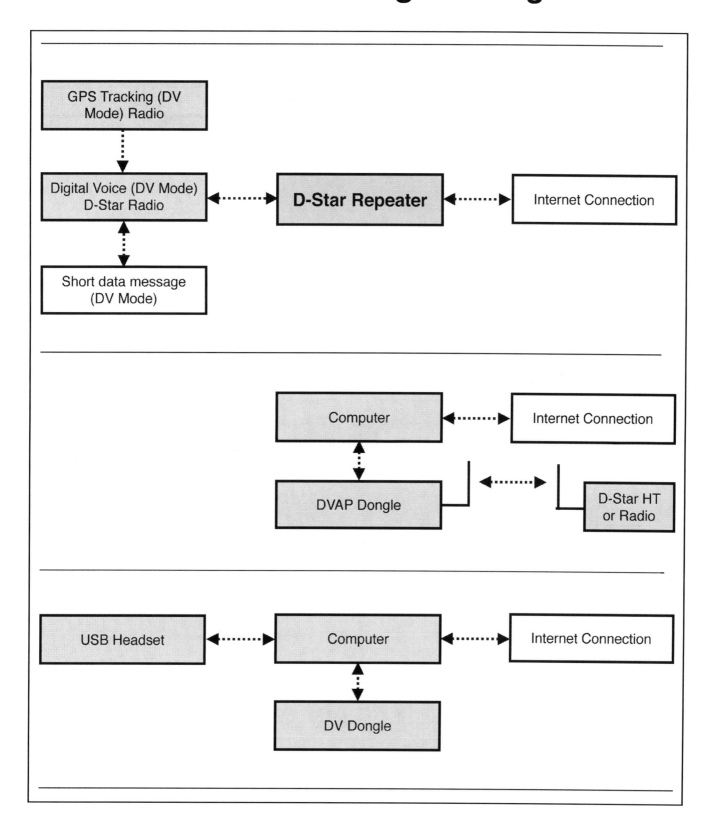

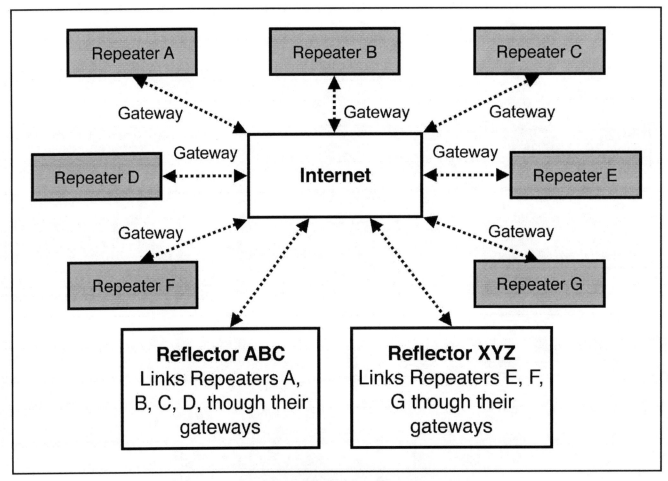

Figure 1 - D-Star Basic Configuration
Figure 2 - D-Star Reflectors and Gateways

D-Star

First, let us look at what an HF radio is. HF radio requires an expensive transceiver, a large antenna, and usually an antenna tuner. Then, one must pray for favorable propagation at specific times of the day or night. Often, contacts are noisy and hard to hear.

Now, let us talk about D-Star. D-Star is a digital voice and text mode, which enables it to make connections via the internet. With a simple VHF/UHF D-Star capable transceiver that costs about the same as most dual band VHF/UHF transceivers, one can talk to the world with crystal clear audio. The antenna is the same VHF/UHF antenna that we all commonly use. If cost is a major factor, a single band transceiver can be used.

Or, a DV Dongle, a Mac or PC computer with an internet connection, and a headset can be used. This is the least expensive and easiest way to get on the air with D-Star.

Or, a DVAP Dongle, a Mac or PC computer with an internet connection, and a D-Star radio can be used.

Once a D-Star connection is established, one can connect to reflectors and repeaters all around the world as seen in Figure 2.

There are repeater maps, repeater lists, and other information at: **http://www.dstarinfo.com**

D-Star Registration Steps

Make sure you register at one site and only one site. Follow ALL of the steps shown at **http://www.dstargateway.org/D-Star_Registration.html** This is a two-step process. After approval, further steps are required. Check your registration at: http://dstar.info/query.html

1. Find a D-Star repeater in your area that has a website that you can register with.

2. Follow the registration steps.

3. Wait for an email from the D-Star administrator approving your registration or try to log in periodically to check on the status of your registration at the same URL you used to register with using your callsign (IN UPPER CASE) and the password you entered during registration. If your registration is pending approval, you will see an Error that the registration has not yet been approved. If it is approved, you will be granted access and you will be able to log into the system and configure your personal information. **THIS NEXT STEP IS REQUIRED**. Once you are logged in, click on "Personal Information" at the right of the page.

4. Click on the checkbox next to the number "1". Then click inside the "Initial" box to the right of your callsign on the same line as the number "1". Type in a single space character. This will not show up but is very important. Do not click on the "RPT" check box. In the "pcname" box, enter your callsign in lower case followed by a dash "-" followed by your type of radio, e.g. 2820 or dvdongle. All characters in the "pcname" box should be lower case and there should be no spaces. When complete, click on "Update".

Some Common D-Star Repeaters

D-Star Repeaters connect to an internet gateway. Reflectors link specific gateways. All linked gateways hear all traffic from all of the gateways connected to the reflector. Reflectors provide a way to link multiple gateways together, providing an easy way to create a wide-area repeater, hold a multi-gateway net, etc. on the fly. Each reflector has three modules, A, B and C.

Some common US reflectors:
- REF069C (New England Repeaters)
- REF030C (Georgia/Southeast Repeaters)
- REF009C (Arizona Permalink Repeaters)
- REF014B (US West Coast Repeaters)

D-STAR (Digital Smart Technologies for Amateur Radio) is a digital voice and data protocol specification developed for use in amateur radio. D-Star compatible radios are available on VHF and UHF amateur radio bands. In addition to the over-the-air protocol, D-Star enables D-Star radios to be connected to the Internet or other networks. D-STAR is the result of research by the Japan Amateur Radio League to investigate digital technologies for amateur radio.

Call Sign Table Setup

To call a person (M1ABC) simplex:
UR: CQCQCQ
RPT1:
RPT2:
Push the PTT and speak: "M1ABC. This is WB2LUA."

To call CQ or a person on a local repeater only:
UR: CQCQCQ
RPT1: W2ABC B (the "B" is in the 8th position with 2 spaces before it)
RPT2: W2ABC G (the "G" is in the 8th position with 2 spaces before it)
Push the PTT and speak: "This is WB2LUA."
Push the PTT and speak: "M1ABC. This is WB2LUA."
If the repeater is not linked, the signal stays locally.

To call CQ on the entire D-star world wide gateway through a local repeater:
UR: CQCQCQ
RPT1: W2ABC B (the "B" is in the 8th position with 2 spaces before it)
RPT2: W2ABC G (the "G" is in the 8th position with 2 spaces before it)
Push the PTT and speak: "CQ CQ CQ. This is WB2LUA in Northport, New York."
If the repeater is not linked, the signal stays locally.

To call CQ on a distant repeater on the D-Star world wide gateway through a local repeater:
UR: /K6ABC B (the "B" is in the 8th position with 2 spaces before it)
RPT1: W2ABC B (the "B" is in the 8th position with 2 spaces before it)
RPT2: W2ABC G (the "G" is in the 8th position with 2 spaces before it)
Push the PTT and speak: "This is WB2LUA in Northport, New York through the W2ABC repeater port B Zone Routing"

To call a person on a distant repeater on the D-Star world wide gateway through a local repeater:
UR: /K6ABC B (the "B" is in the 8th position with 2 spaces before it)
RPT1: W2ABC B (the "B" is in the 8th position with 2 spaces before it)
RPT2: W2ABC G (the "G" is in the 8th position with 2 spaces before it)
Push the PTT and speak: "M1ABC. This is WB2LUA in Northport, New York through the W2ABC repeater port B Zone Routing"
You would need to know if this person was on the repeater.

To call to a person on the D-Star world wide gateway through a local repeater:
UR: M1ABC
RPT1: W2ABC B (the "B" is in the 8th position with 2 spaces before it)
RPT2: W2ABC G (the "G" is in the 8th position with 2 spaces before it)
Push the PTT and speak: "M1ABC. This is WB2LUA in Northport, New York through the W2ABC repeater port B call signRouting"

To connect a local repeater to a DPlus Reflector on the D-Star world wide gateway through:
To link the repeater to a reflector
UR: REF069CL
R1: W2ABC B (the "B" is in the 8th position with 2 spaces before it)
R2: W2ABC G (the "G" is in the 8th position with 2 spaces before it)
Push the PTT to link the repeater to a reflector
Reset as follows to talk
UR: CQCQCQ
R1: W2ABC B (the "B" is in the 8th position with 2 spaces before it)
R2: W2ABC G (the "G" is in the 8th position with 2 spaces before it)
Push the PTT and speak: "This is WB2LUA in Northport, New York
To unlink a repeater from a reflector
UR: U (the "U" is in the 8th position with 7 spaces before it)
R1: W2ABC B (the "B" is in the 8th position with 2 spaces before it)
R2: W2ABC G (the "G" is in the 8th position with 2 spaces before it)
Push the PTT to unlink the repeater from the reflector
Most repeater owners only allow administrators to link the reflector.

Reflectors allow multiple D-Star repeaters and Dongle users, from around the world, to be joined together and whatever information is transmitted across one of the repeaters is repeated across all of the connected repeaters. If you connect to a reflector, you will be able to talk to everyone on every repeater that is connected to that reflector

Some common US reflectors:
REF069C (New England Repeaters)
REF030C (Georgia/Southeast Repeaters)
REF009C (Arizona Permalink Repeaters)
REF014B (US West Coast Repeaters)

A complete listing of reflectors is available at www.dstarinfo.com

Icom ID-5100A Programming

Frequency
1. Touch VFO
2. Touch MHz
3. Touch F-INP for direct input

Repeater Offset (Menu F-3)
1. Press DUP and select the offset

Tone (Menu F-3)
1. Press TONE
2. Select TONE

Tone Frequency
1. Press main MENU
2. Press DUP/TONE
3. Change frequency

Memory In (Analog Menu F-1)
1. Press MW
2. Press WRITE to selected channel
3. Press CH SEL to select the memory channel
4. Press on the channel number
5. Press WRITE

Memory Recall (Analog Menu F-1)
1. Press V/M
2. Select the desired channel

D-Star - Call Sign Setup (one time setup)
1. Press MENU
2. Press My Call Sign
3. Press MY call sign memory channel 1 for 1 second
4. Press Edit
5. Enter call sign
6. Press ENT

D-Star - Entering a Text Message
1. Press MENU
2. Press My Station > TX Message
3. Press desired message memory number for 1 second
4. Press Edit
5. Enter message
6. Press ENT

D-Star Simplex Call
1. Press DR
2. Press FROM, opens FROM SELECT
3. Repeater List, opens REPEATER GROUP
4. Press Simplex
5. Press a desired frequency, which is displayed in FROM and CQCQCQ is displayed in TO

D-Star - Connecting to a Repeater (From)
1. Press DR
2. Press FROM, opens FROM SELECT
3. Touch Repeater List, opens REPEATER GROUP
4. Touch Repeater Group and select the appropriate group

D-Star - Connecting to a Repeater Direct (To)
1. Press TO (while in DR mode)
2. Press what is in the TO field
3. Press Local CQ. CQCQCQ should appear in the To field

D-Star - Connecting to a Reflector through a Repeater (To)
1. Press TO (while in DR mode)
2. Press what is in the TO field
3. Press REFLECTOR
4. Select Use Reflector

D-Star - Check if you can access a repeater
1. Hold down the PTT for 1 second
2. You will get a reply call such as UR? or RPT?

D-Star - DVAP
1. Press DR and set the frequency specified in the DVAP software
2. Press TO
3. Press Direct Input (UR)
4. Input Reflector code, such as REF069CL in the CQ field to link the DVAP to the Reflector. Note the L stands for link. To unlink from the reflector, use U in the last field.
5. Press PTT.
6. Change the UR to CQCQCQ.

Memory In (D-Star)
1. In the DR screen, select the settings to be saved
2. Press the Function group icon to select MENU D3
3. Press MW
3. Press CH SEL to select the memory channel
4. Press on the channel number
5. Press WRITE

Memory Recall (D-Star)
1. Press DR to close the DR Screen
2. Press the Memory Channel Number
3. Press MR
4. Rotate DIAL to select a memory channel

Icom ID-51A Plus

Power Level
1. Press and hold V/MHZ for 1 second to select power level.

Frequency
1. Press V/MHz to select 1 MHz or 10 MHz
2. Rotate DIAL

Squelch Level
1. Hold down SQL and rotate DIAL

Repeater Offset
1. Press QUICK
2. Press DUP
3. Select DUP- or DUP+

Tone
1. Press QUICK
2. Press TONE
3. Select TONE

Tone Frequency
1. Press main MENU
2. Press DUP/TONE
3. Press Repeater Tone
4. Change frequency

Memory In
1. Hold S.MW for 1 second
2. Rotate DIAL to select channel
3. Hold S.MW for 1 second. Three beeps will sound

Memory Recall
1. Press M/CALL, MR appears
2. Rotate DIAL to Select the desired channel

D-Star - Call Sign Setup (one time setup)
1. Press MENU
2. Select My Station
3. Select My Call Sign
4. Press ENT
5. Select MY Call Sign memory channel 1
6. Push QUICK, then select EDIT
7. Enter call sign
8. Press D-Pad (Ent)

D-Star - Entering a Text Message
1. Press MENU
2. Press My Station > TX Message
3. Press desired message memory number for 1 second
4. Press QUICK and Select EDIT
5. Enter message
6. Press ENT twice

D-Star Simplex Call
1. Press DR
2. Press FROM, opens FROM SELECT
3. Repeater List, opens REPEATER GROUP
4. Press Simplex
5. Press a desired frequency, which is displayed in FROM and CQCQCQ is displayed in TO

D-Star - Connecting to a Repeater (From)
1. Press and hold DR for 1 second
2. Select FROM, ENT
3. Select Repeater List, opens REPEATER GROUP
4. Select Repeater Group and select the appropriate group

D-Star - Connecting to a Repeater Direct (To)
1. Press and hold DR for 1 second
2. Press TO, ENT
3. Press Local CQCQCQ

D-Star - Connecting to a Reflector through a Repeater (To)
1. Press TO, ENT
2. Select Reflector
3. Select and use Reflector

D-Star - Check if you can access a repeater
1. Hold down the PTT for 1 second
2. You will get a reply call such as UR? or RPT?

D-Star - DVAP
1. Press DR and set the frequency specified in the DVAP software
2. Set all Reflector and Repeater functions
3. Select MENU
4. Select Call Sign
5. In the UR Field:
6. Input Reflector code, such as REF069CL in the CQ field to link the DVAP to the Reflector. Note the L stands for link. To unlink from the reflector, use U in the last field.
7. Press PTT.
8. Change the UR to Field to CQCQCQ.
9. Press PTT

Memory In (D-Star)
1. Set all DV settings
2. Press V/MHZ
3. Press S.MW for 1 second
4. Rotate DIAL to desired memory channel
5. Hold S.MW for 1 second

Memory Recall (D-Star)
1. Press DR to close the DR Screen
2. Press the Memory Channel Number
3. Press MR
4. Rotate DIAL to select a memory channel

Icom IC-2820H Programming

Selecting the operating frequency band
Push and hold the desired band's (left or right) **[MAIN BAND]** for 1 sec. then rotate the appropriate band's **[DIAL]**.

Tuning Step Selection for extra decimal places
1. Push **[F]** to display the function guide.
2. Push [TS] to enter the tuning step set mode.
3. Rotate [DIAL] to select 12.5 KHz. to obtain the extra decimal places.
4. Push [F] to exit turning steps.
5. Start at 444.2 MHz and go up one click, which will become 444.21250 MHz.

Repeater Offset
1. Set the receive frequency (repeater output frequency) on the main band.
2. Push **[DUP]** one or two times, to select minus duplex or plus duplex (DUP- or DUP+)

Repeater Offset Frequency
1. Push [F]. Then, [MENU].
2. Rotate [DIAL] and select "DUP/TONE"
3. Select "OFFSET Frequency.
4. Push [DIAL] and change the frequency
5. Push [BACK] (Right band's) twice to exit DUP/TONE set mode.

Tuning Steps
1. Push [F] to display the function guide.
2. Press [TS] and change the frequency steps
3. Press [F] again to exit.

Repeater Tone
1. Push [F] to display the function guide.
2. Push [TONE DTMF] several times until "TONE" appears.
3. To turn OFF the sub-audible tone encoder, push [TONE DTMF] several times until no tone indicators appears.

Repeater Tone Frequency
1. Push [F]. Then, [MENU].
2. Rotate the Tuning Dial [DIAL] to select "DUP/TONE" Then push the [MAIN-BAND] key.
3. Rotate [DIAL] to select "REPEATER TONE." Then push the [MAIN-BAND] key.
4. Rotate [DAIL] to select the sub-audible frequency, then push [MAIN-BAND] key.
5. Push [BACK] (Right band's) twice to exit DUP/TONE set mode.

Setting the Mode
1. Push [F] to display the function guide.
2. Push [MODE] until the desired mode is obtained.

Auto Repeater
1. Push [F] to display the function guide.
2. Push [MENU] (Right band's) to enter MENU screen.
3. Rotate [DIAL] to select "SET MODE" then push [MAIN BAND].
4. Rotate [DIAL] to select "AUTO REPEATER" then push [MAIN BAND].
5. Rotate [DIAL] to select the auto repeater setting.
    USA version:
    "RPT1" : Activates duplex only. (default)
    "RPT2" : Activates duplex and tone.
    "OFF" : Auto repeater function is turned OFF.
6. Push [BACK] (Right band's) twice to exit

GPS On-Off
1. Press the [F] button, then the [MENU] key
2. Rotate [DIAL] to select "SET MODE," then push [MAIN BAND].
3. Rotate [DIAL] to select "GPS," then push [MAIN BAND].
4. Rotate [DIAL] to turn GPS on or off, then push [MAIN BAND].
5. Push [BACK] twice to return to frequency indication.

Storing a Memory Channel
1. Push and hold the same band's [M/CALL] for 1 sec., then rotate the same band's [DIAL] to select the desired memory channel. "X" indicator and memory channel number blink.
3. Push and hold [S.MW] (Left band's) for 1 sec. to program. 3 beeps will sound and return to VFO mode automatically after programming.
Store memory after D-Star repeaters are stored so that the frequency and call signs will be stored.

Programming My Call Sign into Memory
1. Press the [F] button, then the [MENU] key
2. Push [MENU] to enter MENU screen.
3. Rotate [DIAL] to select "CALL SIGN MEMORY." Then, push [MAIN BAND].
4. Rotate [DIAL] to select "MY CALL SIGN MEMORY" then push [MAIN BAND].
5. Rotate [DIAL] to select the desired call sign channel (M01 to M06), then push [MAIN BAND].
6. Rotate [DIAL] to select the desired character, then push [>] to move the cursor right. Push [<] to move the cursor left.
    Up to 8-character call signs can be entered.
    Push [ABC] to select the character group from capital letter characters.
    Push [12/] to select the character group from numbers or symbols.
    Push [CLR] to clear the selected character.
7. Repeat step 6 until your own call sign is programmed.
8. Push [>] several times to move the cursor to "/" position.
    A to Z, 0 to 9 and "/" characters are available.
    Example: WB2LUA /John
9. Repeat step y to program the desired 4-character note.
10. Push [MAIN BAND] to store the programmed call sign.
11. Push [BACK] three times to return to frequency indication.

Programming Transmit Message into Memory

1. Push [F] to display the function guide.
2. Push [MENU] to enter MENU screen.
3. Rotate [DIAL] to select "DV MESSAGE," then push [MAIN BAND].
4. Rotate [DIAL] to select "TX MESSAGE MEMORY," then push [MAIN BAND].
5. Rotate [DIAL] to select the desired message memory channel, 01 to 05, then push [MAIN BAND]. Previously message is displayed if programmed.
6. Rotate [DIAL] to select the desired character.
    Push [Aa] to turn the character group from alphabetical characters capital letters or lower case letters.
    Push [1/] (Right band's) to turn the character group from numbers or symbols.
    Push [>] or [<] (Left band's) to move the cursor right or left, respectively.
    Push [CLR] to clear the selected character.
    example: IC-2820H, John in Northport, NY
7. Repeat the step y to enter the desired message.
    Up to 20-character messages can be set.
8. Push [MAIN BAND] to store the message.
9. Push [BACK] (Right band's) twice to exit from DV message screen.

Programming D-Star Stations into Memory

1. Push [F] to display the function guide.
2. Push [MENU] to enter MENU screen.
3. Rotate [DIAL] to select "CALL SIGN MEMORY" then push [MAIN BAND].
4.. Rotate [DIAL] to select "YOUR CALL SIGN MEMORY." Then push [MAIN BAND].
5. Rotate [DIAL] to select the desired call sign channel (U01 to Uxx), then push [MAIN BAND].
6. Rotate [DIAL] to select the desired character, then push [>] to move the cursor right. Push [<] to
    move the cursor left.
    Up to 8-character call signs can be entered.
    Push [ABC] to select the character group from capital letter characters.
    Push [12/] to select the character group from numbers or symbol in the 8th position.
    ("C" 144 MHz. "B" 440 MHz.) Example: W9ABC B
    Push [CLR] to clear the selected character.
    Usually "YOUR CALL" is set to CQCQCQ
7. Repeat step 6 until the desired station call sign is programmed.
8. Push [MAIN BAND] to store the programmed call sign.
9. Push [BACK] three times to return to frequency indication.

Programming D-Star Repeaters into Memory

1. Push [F] to display the function guide.
2. Push [MENU] to enter MENU screen.
3. Rotate [DIAL] to select "CALL SIGN MEMORY" then push [MAIN BAND].
4.. Rotate [DIAL] to select "RPT1." Then push [MAIN BAND].
5. Rotate [DIAL] to select the desired call sign channel, (R01 to Rxx) then push [MAIN BAND].
6. Rotate [DIAL] to select the desired character, then push [>] to move the cursor right. Push [<] to move the cursor left.
    Up to 8-character call signs can be entered.
    Push [ABC] to select the character group from capital letter characters.
    Push [12/] to select the character group from numbers or symbol in the 8th position.
    ("C" 144 MHz. "B" 440 MHz.) Example: W9ABC B
    Push [CLR] to clear the selected character.
7. Repeat step 6 until the desired station call sign is programmed.
8. Push [MAIN BAND] to store the programmed call sign.
9. Push [BACK] three times to return to frequency indication.
10. Repeat steps 1 to 9 for each repeater that is to be programmed into memory.

Call Sign Table Setup (When Ready to Transmit)

1. Push [F] twice to display the function guide.
2. Push [CS] to display the "CALL SIGN" screen.
3. Rotate [DIAL] to select "RPT1," then push [MAIN BAND].
    RPT1 CALL SIGN screen is displayed.
4. Rotate [DIAL] to select the local repeater's call sign, then push [BACK].
5. Rotate [DIAL] to select "RPT2" then push [MAIN BAND].
    RPT2 CALL SIGN screen is displayed.
6. Rotate [DIAL] to select the desired repeater's call sign.
7. Push [BACK] to exit "CALL SIGN" screen or remain in this screen.

Call Sign Squelch

Set the desired operating frequency in DV mode, Digital code and MY CALL SIGN.
1. Push [TONE DTMF] several times to activate the digital code or digital call sign squelch (DSQL or CSQL) Digital call sign squelch "DSQL," Digital call sign beep "DSQL ," Digital code squelch "CSQL," Digital code beep "CSQL" and no tone operation are activated in order.
2. Operate the transceiver in the normal way.
3. When the received signal includes a matching call sign/code, the squelch opens and the signal can be heard. When the received signal's call sign/code does not match, digital call sign/digital code squelch does not open; however, the S/RF sign/code, the squelch opens and the signal can be heard. When the received signal's call sign/code does not match, digital call sign/digital code squelch does not open; however, the S/RF indicator shows signal strength.

Received Call Sign

When a call is received in DV mode, the calling station and the repeater call signs being used can be stored into the received call record. The stored call signs are viewable in the following manner. Up to 20 calls can be recorded.

Call Record
1. Display the RX call sign record screen;
    Accessing from MENU screen:
        A. Push [F] to display the function guide.
        B. Push [MENU] to display the "MENU" screen.
        c. Rotate [DIAL] to select "RX CALL SIGN," then push [MAIN BAND].
    Accessing from function guide 2:
        A. Push [F] twice to display the function guide 2.
        B. Push [CD] to display the "RX CALL SIGN" screen.
2. Rotate [DIAL] to select the desired record.
3. Push [MAIN·BAND] to display the received call details.
    CALLER   The station call sign that made a call
    CALLED   The station call sign called by the caller.
    RXRPT1   The repeater call sign used by the caller station.
    RXRPT2   The repeater call sign linked from RXRPT1.
4. Push [MAIN BAND] or [BACK] to return to the "RX CALL SIGN" screen.
5. Push [BACK] to exit from the "RX CALL SIGN" screen.

One-touch Call Reply
1. After receiving a call, push [F] twice to display the unction guide 2.
2. Push [R>CS] to set the received call sign a function guide 2. for the call.
    When selecting a call record via MENU screen:
    A. Push [MAIN·BAND] to display the call record details.
    B. Push [R>CS] to set the received call sign to that of the call record.
    Setting from function guide 2:
    Push [R>CS] to set the received call sign to hat of the call record.
3. Push [PTT] to transmit; release to receive.
    Set your own call sign (MY) in advance.
    The call sign stored in "CALLER" is stored as "YOUR," "RXRPT1" is stored as "RPT2" and RXRPT2" is stored as "RPT1."
    Error beeps sound when a call sign is received incorrectly, and no call sign is set in this case.

Low-speed data communication

Turn OFF the GPS data communication to operate the low-speed data communication.

Connection:
Connect the transceiver to your PC using with the cable OPC-1529R

Low-speed data communication application settings:
Port: The same COM port number as IC-2820H's
Baud rate: 9600 bps (fixed value)
Data: 8 bit
Parity: None
Stop : 1 bit
Flow control : Xon/Xoff

Low-speed data communication operation:
Confirm that in AUTO, the computer controls when [PTT] is active, so that you can send data without pressing [PTT] on the radio.
1. Set your own, station call signs, etc. as described in "Digital voice mode operation" and "Digital repeater operation."
2. Refer to the instructions for the low-speed data communication application.
3. To transmit data: At the same time as voice audio, push and hold [PTT] to transmit while sending data from the PC. Release [PTT] to receive.

Transmission condition settings:
1. Push [F] to display the function guide.
2. Push [MENU] to display the "MENU" screen.
3. Rotate [DIAL] to select "DV SET MODE," then push [MAIN BAND].
4. Rotate [DIAL] to select "DV DATA TX," then push MAIN BAND].
5. Rotate [DIAL] to select the desired transmission condition.
   PTT: The entered text data in the Terminal Window (buffer screen) is transmitted when [PTT] is pushed. (default)
   AUTO : The entered text data in the Terminal Window (buffer screen) is automatically transmitted when text is entered.
6. Push [BACK] three times to exit from DV set mode screen.

Icom IC-92AD Programming

Selecting the operating frequency band
Push [MAIN.DUAL] to toggle between A and B band.

Setting the Squelch Level
While pushing and holding [SQL], rotate [DIAL] to select squelch level.

Tuning Step Selection for extra decimal places
1. Push [VFO] to select VFO mode.
2. Push [BAND] to select the desired frequency band.
3. Push [TS] for 1 second to display frequency steps.
4. Rotate [DIAL] to select 12.5 KHz. to obtain the extra decimal places.
5. Push [TS] for 1 second to store turning steps and exit

Repeater Offset
Push [DUP] for 1 second to select offset (DUP- or DUP+).

Repeater Offset Frequency
1. Push [MENU].
2. Rotate [DIAL] and select "DUP/TONE", then [5,]
3. Select "OFFSET Frequency, then [5,]
4. Rotate [DIAL] and change the frequency
5. Push [MENU] to exit.

Repeater Tone
Push [TONE] for 1 second to activate tone.

Repeater Tone Frequency
1. Push MENU
2. Rotate [DIAL] to select DUP/TONE, then [5,]
3. Rotate [DIAL] to select RPT TONE, then [5,]
4. Rotate [DIAL] to select the desired tone frequency.
5. Push [MENU] to exit.

Setting the Mode
Push and hold [MODE] for 1 second several times to select the desired mode. MY CALL should be programmed first to enter the DV mode.

Auto Repeater
1. Push MENU
2. Rotate [DIAL] to select AUTO RPT, then [5,]
3. Rotate [DIAL] to select:
    "RPT1" : Activates duplex only. (default), then [5,]
    "RPT2" : Activates duplex and tone, then [5,]
    "OFF" : Auto repeater function is turned OFF, then [5,]

Storing a Memory Channel
1. Push and hold the same band's [S.MW] for 1 sec., then rotate the [DIAL] to select the desired memory channel.
3. Push and hold [S.MW] for 1 sec. to program.
Store memory after D-Star repeaters are stored so that the frequency and call signs will be stored.

Programming My Call Sign into Memory
1. Select B band by pushing [MAIN/DUAL].
2. Push MENU
3. Rotate [DIAL] to select CALL SIGN, then [5,]
4. Rotate [DIAL] to select MY, then [5,] and CALL SIGN screen is displayed
5. Rotate [DIAL] to select memory M01 to M06
6. Push > to enter programming mode.
7. Rotate [DIAL] to select the desired character. (Push A/a to change characters).
8. Push <> to select the next or previous digit.
    Example: WB2LUA /John
9. Push enter to store the call sign.
10. Push [MENU] to return to frequency indication.

Programming Transmit Message into Memory
1. Select B band.
2. Push MENU
3. Rotate [DIAL] to select MESSAGE/POSITION, then [5,]
4. Rotate [DIAL] to select TX MESSAGE
5. Rotate [DIAL] to select memory CH01 to CH06
6. Push > to select message edit.
7. Push > to enter programming mode.
8. Rotate [DIAL] to select the desired character. (Push A/a to change characters).
9. Push <> to select the next or previous digit.
    Example: 92,John,Northport, NY
10. Push [5] to store the call sign.
11. Push [MENU] to return to frequency indication.

Programming D-Star Stations into Memory
1. Select B band.
2. Push MENU
3. Rotate [DIAL] to select CALL SIGN, then [5,]
4. Rotate [DIAL] to select UR, then [5,] and YOUR CALL SIGN screen is displayed
5. Rotate [DIAL] to select memory U01 to U60
6. Push > to enter programming mode.
7. Rotate [DIAL] to select the desired character. (Push A/a to change characters).
    Usually "YOUR CALL" is set to CQCQCQ
8. Push enter to store the call sign.
9. Push [MENU] to return to frequency indication.

Programming D-Star Repeaters into Memory
1. Select B band.
2. Push MENU
3. Rotate [DIAL] to select CALL SIGN
4. Rotate [DIAL] to select R1/R2
5. Push [MENU] and RPT1 or RPT2 CALL SIGN screen is displayed
6. Rotate [DIAL] to select memory R01 to R60
7. Push > to enter programming mode.
8. Rotate [DIAL] to select the desired character. (Push A/a to change characters).
    Example: W9ABC B
9. Push enter to store the call sign.
10. Push [MENU] to return to frequency indication.

Call Sign Table Setup (When Ready to Transmit)
Set MY CALL (MY):
1. Set the desired frequency in B band.
2. Push MENU
3. Rotate [DIAL] to select CALL SIGN, then [5,]
4. Rotate [DIAL] to select MY, then [5,]
5. Rotate [DIAL] to select your own programmed call sign.
6. Push [5].

Set Station Call Sign or CQ (UR):
1. Set the desired frequency in B band.
2. Push MENU
3. Rotate [DIAL] to select CALL SIGN, then [5,]
4. Rotate [DIAL] to select UR, then [5,]
5. Rotate [DIAL] to select a programmed call sign.
6. Push [5].

Set Repeater Call Sign (R1 and R2)
1. Set the desired frequency in B band.
2. Push MENU
3. Rotate [DIAL] to select CALL , then [5,]
4. Rotate [DIAL] to select R1 or R2, then [5,]
5. Rotate [DIAL] to select a programmed repeater call sign.
6. Push [5].

DVAP Dongle Programming

The DVAP Dongle is used with a D-Star HT or other D-Star radio. It is essentially a mini D-Star repeater. Connect the DVAP to the USB port of the computer.

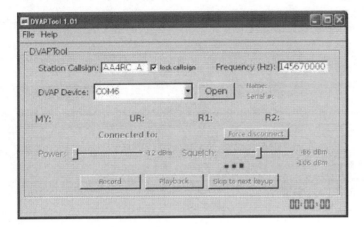

Registration

The first step is to GET REGISTERED! If you are not already registered on the D-STAR network, you need to get registered. It is very important that you register with ONE gateway only. Once you are registered on one gateway, you are registered on them all. Follow the instructions at http://www.dstargateway.org for registration. Until you are registered, you will only be able to listen to connected systems. No transmission is allowed for unregistered call signs.

It is recommended that you go to the gateway where you registered and add a terminal entry with an "A" in the initial field (it may take several hours for the update to propagate to all gateways).

Computer Setup

Next, make sure you are connected to the Internet. All communications between the DVAP (Digital Voice Access Point) and a gateway or reflector use your Internet connection.

To start using the DV Access Point Dongle, double click on the DVAPTool icon on your desktop. DVAPTool will start and display a dialog window similar to the picture below.

There are two text entry boxes that require input, the station call sign and the Frequency box. Enter your personal callsign in the Station call sign text box. Use the "A" in the eighth character of your call sign as shown above.

Enter a simplex frequency for operation in the Frequency box. Consult the frequency band plan for your country/area for appropriate frequencies. Enter the entire frequency in Hz (e.g. 146550000 for 146.550 MHz).

The "lock call sign" checkbox next to your Station call sign will allow you to limit the radio users of your DVAP to only your call sign or any call sign. This is the way you can prevent others from using your DVAP.

The DVAP Device drop down selection lists devices that use the FTDI chip used in the DV Access Point Dongle device and other serial to USB devices. Select the appropriate device.

Click on the "Open" button to connect to the DV Access Point Dongle. (www.dvapdongle.com)

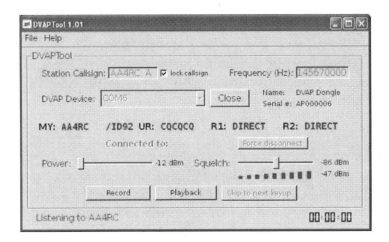

Once the DV Access Point Dongle is opened, the Device Name and Serial Number fields should show data specific to the hardware you have opened. If it does not, you may need to select a different device in the DVAP Device box.

DVAPTool then makes a request to an Internet based D-STAR gateway name server. If DVAPTool has problems connecting to the server, a message box will pop up letting you know that there was a problem. Check your Internet connection and try again.

Set the Power level slider to maximum (10dBm or 10mW) and set the Squelch level slider to half way between your current noise level (the red bars below the slider) and the max level (similar to the pic above).

Record and playback conversations by clicking the "Record" and "Playback" buttons. While playing back, you can click the "Skip to next keyup" to skip to the next person talking. Note that playback is done only to your radio, not to a connected gateway or reflector.

You can walk away from the PC/Mac now since all control commands are issued from your D-STAR radio.

The DV Access Point Dongle has four LED's which indicate the current operating status. They are colored green, red, yellow, and blue (in that order from left to right).

The green LED shows the mode of operation, slow pulsing indicating idle and solid green shows the DVAP is receiving RF from a radio.

The red LED shows that the DVAP is transmitting to your radio.

The yellow LED shows data under runs from the connected gateway/reflector or missed D-STAR packets from your radio.

The blue LED shows data is being sent from the PC/Mac to the DVAP.

Status Indicators Operating Notes

The DV Access Point Dongle is a high speed, real time device. It communicates with the PC/Mac at 230Kbps and needs adequate CPU speed and time to operate properly. Many operations on the PC/Mac can interfere with normal operations. These include screen savers, web browsers, instant messengers, etc. For best operation, avoid running CPU intensive applications when operating the DV Access Point Dongle.

To start using the DV Access Point Dongle, double click on the DVAPTool icon on your desktop. DVAPTool will start and display a dialog window similar to the picture below.

There are two text entry boxes that require input, the station call sign and the Frequency box. Enter your personal call sign in the station call sign text box. It is recommended (but not required) that you add an "A" terminal in your gateway registration and that you use the "A" in the eighth character of your call sign as shown below.

Enter a simplex frequency for operation in the Frequency box. Consult the frequency band plan for your country/area for appropriate frequencies. Enter the entire frequency in Hz (e.g. 146550000 for 146.550 MHz).

The "lock call sign" checkbox next to your station call sign will allow you to limit the radio users of your DVAP to only your call sign or any call sign. This is the way you can prevent others from using your DVAP.

The DVAP Device drop down selection lists devices that use the FTDI chip used in the DV Access Point Dongle device and other serial to USB devices. Select the appropriate device.

Click on the "Open" button to connect to the DV Access Point Dongle.

Enter your call sign into the MYCALL field of your radio. Select DV mode and configure simplex operation, usually by holding down the "DUP" key until there is no "dup+" or "dup-" on the radio display. You do not need to enter RPT1 nor RPT2 since the radio will place "DIRECT" in both when in simplex mode. All commands are entered into the URCALL field as detailed below. Enter the following values in the URCALL field on your radio:

"DVAP   I" (request voice ID from the DVAP)

"DVAP   E" (key down and speak for echo test)

"CQCQCQ "(transmit to a connected system)

"xxxxxxmL" (to link, replace XXXXXX with the gateway or reflector call sign making sure to use 6 characters filling the end with spaces as needed. Replace the "m" in the 7th character with the module to which you wish to link. Use "L" in the 8th character to indicate the Link command. For example, to link to Reflector 001 module C, use "REF001CL" in URCALL. To link to gateway W4DOC module A, use "W4DOC AL" in URCALL)

"   U" (unlink from a linked gateway/reflector. Make sure the "U" is in the eighth position of URCALL)

Note that, other than "CQCQCQ ", the commands above are processed locally and not transmitted to a connected system.

Radio Command Reference Customizing Voice Message Playback

You can replace the included files that are played to indicate command actions. The files are:

alreadylinked.dvrec (played when you try to link when linked)

alreadyunlinked.dvrec (played when you try to unlink when unlinked)

dvap-id.dvrec (played when "DVAP   I" is issued)

gatewayunknown.dvrec (played when the requested gateway is unknown)

remotesystemliked.dvrec (played when the requested system is linked)

remotesystemlinkedro.dvrec (played when the requested system is linked receive only)

remotesystemunlinked.dvrec (played when the linked system is unlinked)

To record a file, make sure the DVAP is open and not connected to a remote system. Click the "Record" button. Key down and speak your announcement. Click "Stop recording". Listen to your recording by clicking on "Playback". You may record the announcement as many times as you like until you are happy with it. DVAPTool places the recording in the file dvaptool.dvrec in the folder from which it was started. Simply rename "dvaptool.dvrec" to one of the above file names to have DVAPTool use it instead of the default.

Support and Updates

Please make sure to join the DVAPDongle yahoo group for program update announcements. Go to http://groups.yahoo.com/group/DVAPDongle to join.

Send support questions/issues or suggestions to support@dvapdongle.com

Connecting and Talking

It is said that reflectors have better bandwidth than the individual repeaters. Therefore, it is better to connect to a reflector than to an individual repeater.

Set your radio to low power so the dongle is not overloaded.

Step 1 - Set UR to a reflector and push the PTT once to establish the link. For example, REF001C. The DV Tool program will show the link and the reflector should return an audio saying "Link Established."

Step 2 - Set UR to CQCQCQ and talk. The DV Tool should show "transmitting to gateway" at the bottom of the window.

Some common US reflectors:
REF069C (New England Repeaters)
REF030C (Georgia/Southeast Repeaters)
REF009C (Arizona Permalink Repeaters)
REF014B (US West Coast Repeaters)

Testing

Connect to the remote system "E" module to run an echo test completely through the network. Connect to "REF030EL" and key up and talk for a few seconds. The audio and data should appear on the screen in a few seconds.

DV Dongle Programming

A DV Dongle is a D-star device that is used with a computer that is connected to the internet and a headset. Connect the DV Dongle to the USB port of the computer and the headset to the audio jacks. After much trial and error, its seems that the sound card headsets don't work well with the DV Tool software. I got very poor audio reports. USB headsets works well with the DV Tool Software. The one recommended is the Logitech USB Comfort headset.

Download the latest **DV Tool** software from **www.dvdongle.com**

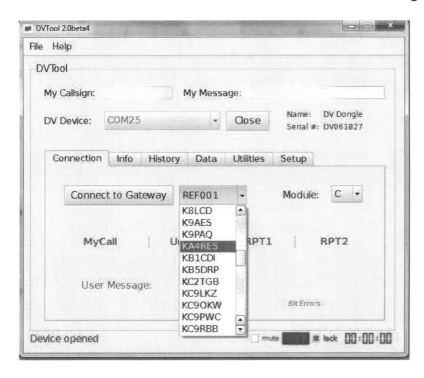

Connect the headset to a USB port.

Install the DV Dongle .exe software. To start using the DV Dongle, double click on the file DV Tool icon. DVTool will start and display a dialog windows similar to the picture above.

Under setup, choose the USB headset audio device. The Audio Input and Output drop down lists allow for the selection of multiple audio devices if they exist.

In Windows control panel or settings for Mac, select sound. Then, adjust the USB microphone sensitivity. Maximum should work well.

Adjust the speaker volume to what is comfortable in the headset. Medium is a setting to start with.

Enter your call sign and message at the top of the screen.

Click on "Open" to connect to the DV Dongle.

Once the DV Dongle is opened, the Device Name, Serial Number, Boot Version, and Firmware Version fields should show data specific to the hardware you have opened.

DVTool then makes a request to an internet based D-Star gateway name server. The D-Star gateways that are available for DVTool connections are listed in the drop down selection box next to "Connect to Gateway".
Connect to a D-Star gateway for receiving and/or transmitting voice. The gateway drop down selection box lists all gateways that are currently running the required software and have IP ports forwarded correctly.

The PTT button (green box in the lower right corner) behaves similar to the PTT button on a radio. When you click and hold, your mic audio is transmitted to the gateway/module selected. When you unclick, the transmission ends.

The DV Dongle has four LED's which indicate the current operating status. The blue LED shows data is being transmitted from the PC/Mac to the device. The yellow LED shows data is being transmitted from the device to the PC/Mac. The green LED shows the mode of operation, slow pulsing indicates idle and fast blinking indicates running. The red LED shows overruns or underruns between the PC/Mac and the device and should normally be off. If you notice frequent red LED activity, your PC/Mac may not be sufficiently fast to operate with the device or you may have other programs running that are taking CPU cycles away from the DVTool application.

The DV Dongle is a high speed, real time device. It communicates with the PC/Mac at 230Kbps and needs adequate CPU speed and time to operate properly. Many operations on the PC/Mac can interfere with normal operations. These include screen savers, web browsers, instant messengers, etc. For best operation, avoid running CPU intensive applications when operating the DV Dongle.

Some common US reflectors:
    REF069C (New England Repeaters)
    REF030C (Georgia/Southeast Repeaters)
    REF009C (Arizona Permalink Repeaters)
    REF014B (US West Coast Repeaters)

Testing

Connect to the remote system "E" module to run an echo test completely through the network. Connect to "REF030EL" and key up and talk for a few seconds. The audio and data should appear on the screen in a few seconds.

Narrow-Band Emergency Message System (EBEMS) - MT-63 Operating Instructions

Narrow Band Emergency Messaging Software (NBEMS) is a system that allows amateur radio operators to send and receive data using almost any computer (Windows, Mac, and Linux) and any analog radio without requiring a dedicated digital infrastructure like packet, D-Star, etc. NBEMS works on VHF/UHF FM and on HF. NBEMS software was developed by Dave Freese - W1HKJ, Stelios Bounanos - M0GLD, Leigh Klotz - WA5ZNU, Stephane Fillod - F8CFE, John Douyere - VK2ETA, Joe Veldhuis - N8FQ, Chris Sylvain - KB3CS, and Gary Robinson - WB8ROL. The three most common modes used are:

MT-63 2000L (long interleave) Digital Mode VHF/UHF FM

MT-63 1000L (long interleave) Digital Mode VHF/UHF FM

MT-63 1000L (long interleave) Mode HF USB

Installation

Download FLDIGI and FLMGS from http://www.w1hkj.com/download.html for Windows, Linux, or MAC.

1. If using a Sound Card Interface, hook up all cables and test the microphone on FM.

2. Connect the RS232 connection from the rig to the computer.

3. Install Fldigi and Flmsg software.

4. Open Fldigi.

5. Click on Configure / Operator and enter your personal information or as much as you want to. Save.

6. Run Sound Card Calibration (CheckSR.exe) and write down the PPM values. Skip this step if using a mac.

    ### 6a. Sound Card Calibration

    1. Download and save CheckSR.exe. It provides the capability of analyzing your sound card offsets and gives you the corrections in parts per million (ppm): http://www.pa-sitrep.com/checksr/CheckSR.exe

2. Open FLDIGI, go to configure, sound card, audio devices tab and make sure you have the sound card you use for your interface properly selected from the capture and playback drop down choices.

3. Under the audio settings tab, you should see a sample rate drop down box for capture and playback. Under each drop down box, select the sample rate that has (native) listed after it and write down this figure. Click save config, then click save. Close FLDIGI.

4. Open CheckSR. From the drop down boxes for sound card settings, Input and Output, choose the sound card you are using. Next, select the sample rate from the drop down box in CheckSR for the sample rate that FLDIGI showed as "Native" then click start.

5. Let the application run for about 5 minutes. You will notice that the numbers will progressively stabilize. After about 5 minutes, click stop then write down the resulting figures on input and output for the measurements in Hz and PPM. Keep this record.

6. Open FLDIGI, go to configure, defaults, sound card and click on the audio settings tab. Enter the PPM figures for RX ppm (CheckSR ppm Input figure) and TX ppm (CheckSR ppm Output figure). If you had a figure that resulted in a minus from CheckSR, enter the PPM setting with the minus symbol followed directly by the figure with no space. Then click save config, then close.

Some sound card softwares have a programmable filter on the mic/line input that can characterize the digital signal as noise, and squash it after 2 seconds. Turned off this filter.

7. Click on Configure / Audio / Devices and select the sound card input and output fields and enter the PPM values from the Sound Card Calibration. Enter the PPM figures for RX ppm (CheckSR ppm Input figure) and TX ppm (CheckSR ppm Output figure). If you had a figure that resulted in a minus from CheckSR, enter the PPM setting with the minus symbol followed directly by the figure with no space. Save.

8. Click on Rig / Rig / Hardware PTT and set up Rig Control. The simplest rig control is to control the push to talk. Set this type of control on the first configuration tab for rig control. Select this operation by checking the "Use serial port PTT". Select the serial port from the list (fldigi will have searched for available ports) or go to Start / Control Panel / System / Hardware / Device Manager / Ports (Com and LPT) to find the com port used. Then specify whether the h/w uses RTS or DTR and whether a + or - voltage is required to toggle PTT on. Then, press the Initialize button. Set the rig to a frequency that is not used during testing. Save.

9. Click Configure /Misc / NBEMS, check: enable, open message folder, open with FLMSG, open in browser, and press FLMSG button and select the executable file. Save.

10. Click Configure / Modems / MT-63 tab, check: 8 bit extended character (UTF-8) and Long Receive Integration. Save.

11. Open FLMSG

12. Configure, Date and time. Select the format you want to use. Save.

13. Configure, Personal Data. This should already be in the program. If not, enter it. Save.

14. Configure Files Formatting. Check open folder when exporting, Callsign, Data-time, and Serial #. Save.

15. From the Form menu, select Radiogram.

16. On the bottom of the window, select Base 64 and MT63-2L from the drop down lists.

Operation

A hard wired sound card interface can be used. Alternately placing the computer's microphone near the radio's speaker to receive and holding the radio's microphone next to the computer speaker while pressing the microphone's PTT switch and pressing PTT on the computer to transmit.

Below are some basic instructions. More advanced instructions can be obtained by watching the instructional videos.

1. Open FLDIGI.

2. Under, Op Mode, select the operating mode MT63-2000L or MT63-1000L (PSK can also be used for other applications)

3. Adjust your sound card master audio volume and sound recording volume to about 75%. You can change this later if necessary according to signal reports from others.

4. In the frequency display, click on the last digit and set it to zero. Then, press enter. Enter the frequency in KHz. on the keyboard. Then, press enter or Select a frequency from the drop down box located after the words "Enter Xcvr Freq."

5. If a sound card interface is not used, place the computer's microphone near the radio's speaker.

6. Open the Squelch button on the bottom right corner of the screen (no color seen in box).

7. Turn the rig squelch off.

8. Set the volume of the radio until the black diamond that changes color to green. Adjust it again on an incoming signal.

9. When there is a solid line in the waterfall, point the cursor to it and click so that a message can be received.

10. Click on the AFC button to track drifting signals.

11a. TO SEND A TEST TONE

    1. Set the signal markers to be centered on the waterfall at 1,500 Hz.

    2. Press the TUNE button (a 1,000 Hz. tone will be transmitted.

    3. Press it again after 15 seconds to release it.

11b. TO SEND A FORMATED MESSAGE:

    1. Open FLMSG

    2. Select the message format you want to use, such as Radiogram or ICS-213.

    3. Enter the appropriate information into the message, including Base 64 and MT63-2KL at the bottom of the window.

    4a. If a sound card interface is not used, place the radio's microphone near the computer's speaker. Press the PTT switch on the microphone, then click "Auto Send."

    4b. If a sound card interface is used, Press the PTT switch on the microphone, then click "Auto Send."

    4c. If a sound card interface and rig control is used, click "Auto Send."

11c. TO SEND AN UNFORMATED MESSAGE:

    1a. If a sound card interface is not used, place the radio's microphone near the computer's speaker. Press the PTT switch on the microphone, then click on the CQ button (green button) or the TTX. To transmit a message that has been typed in, press the PTT switch on the microphone, then click on the TTX button.

    1b. If a sound card interface is used, click on the CQ button (green button) or the TTX. To transmit a message that has been typed in, click on the TTX button.

    1c. If a sound card interface and rig control is used, click on the CQ button (green button) or the TTX button.

FLDI User Manual: http://www.w1hkj.com/FldigiHelp-3.22/operating_page.html

Macros

1. Open Flidigi

2. On the right, above the waterfall, next to the blue blank button is a #1. Left click this number until it changes to "3" or "4", whichever bank you want to use.

3. On the left side, there is a green blank macro button, right click a button and the macro editor will pop up.

4. Copy the line from below. Paste this line into "macro text" box.
DE <MYCALL> ALL STATIONS THIS IS A MESSAGE FROM, Northport, New York [MT63-2KL 1500Hz] DE <MYCALL> K <RX>

5. Enter a button name below into the "macro button label" box (such as TX-30s)

6. Click Apply and Close

7. Choose File, Macros, Save, and Save.

Test it the first time by left clicking it without a radio connected.

Instructional Videos

http://www.youtube.com/watch?v=-1wZ7uIA-Qs

http://www.youtube.com/watch?v=SWZ2vKWSilE&list=PLBF8CFBA57CC6C2CC

http://www.youtube.com/watch?v=psP489NOkg0

http://www.youtube.com/watch?v=9Pw9XWnwukc

http://www.nyc-arecs.org/narrow.html

Echolink and IRLP

Echolink

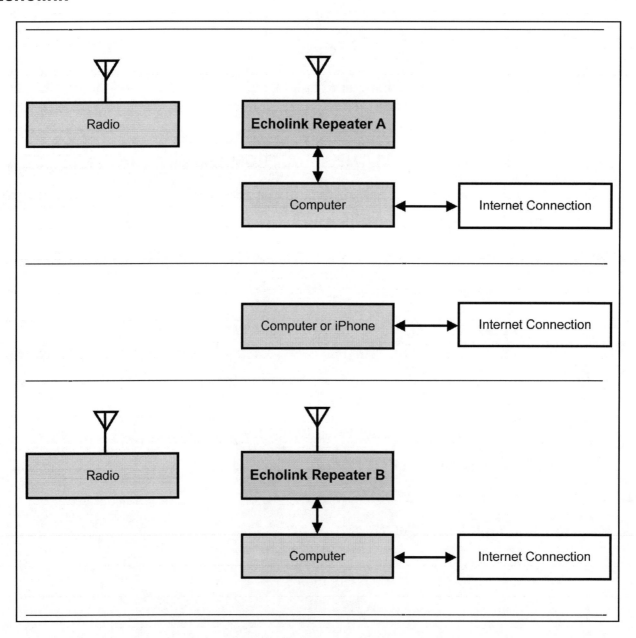

EchoLink makes it possible for amateur radio stations to connect to a repeater anywhere in the world via the internet. The technology uses streaming audio. Echolink was developed by Jonathan Taylor, K1RFD, in early 2002. Connections to be made between Echolink connected; stations, or from computer to station, or from an iPhone to a station or repeater.

The first step is to register your callsign at www.echolink.org to be authenticated. Then, download the PC software from www.echolink.org. Install the software in your computer or iPhone. A list of repeaters will appear on the computer or iPhone screen. Software for the Mac is available from http://echomac.sourceforge.net.

W1AW (ARRL) is also present on the list of active EchoLink stations. However, W1AW is listed as "BUSY" since this connection is used for the conference server. Therefore, DO NOT connect to W1AW. To listen, you must connect to W1AWBDCT (node 501433).

The Apple Airport series router does not support port triggering. It will need to be configured manually to forward to a Mac's IP address.

Step 1. On the mack, go to System Preferences / Network / Advanced / TCP/IP - and find the IPv4 address. On the PC, go to Control Panel / Network Connections. Double click on Network Connections. Click on Support or Details and copy the IP Address. **NOTE: each computer used, will have a different IPv4 address. Consequently, if computers are changed, the IPv4 address must be change in the router. Only one IPv4 address can be used in the router. The iPhone appears to work with any configuration.**

Step 2. Open Airport Utility.

Step 3 . Go to the network tab, Select Port Mapping, click the + sign to add a port mapping and name it EchoLink.

Step 4. Add Public UDP ports: 5198, 5199

Step 5. Add Private UDP Ports: 5198, 5199

Step 6. Add Public TCP Ports: 5200

Step 7. Add Private TCP Ports: 5200

Step 8. Change the Private IP Address to the IPv4 address found in Step 1.

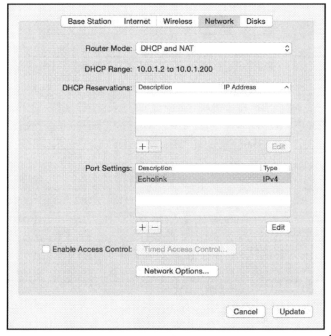
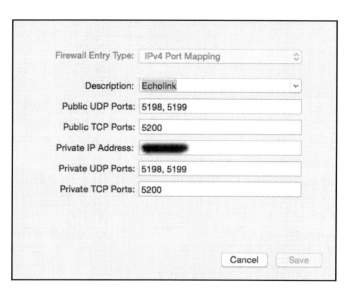

IRLP (Internet Linking Project)

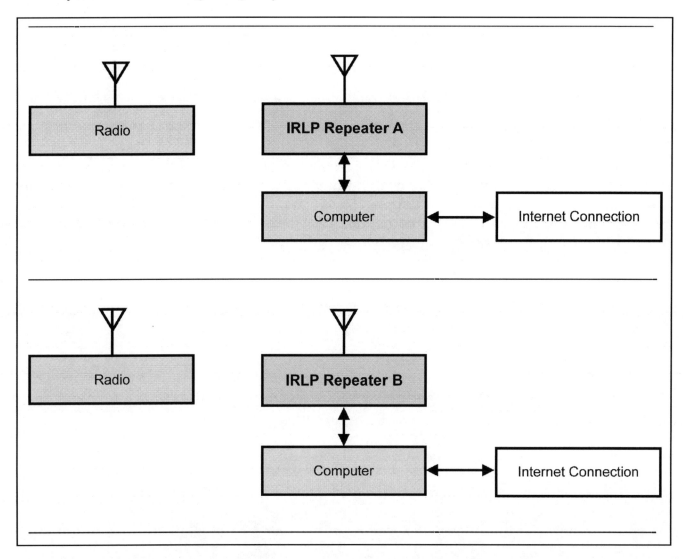

IRLP is a internet-connected collection of radios and repeaters that was invented by David Cameron VE7LTD in Canada in 1997. The official information web page is at http://www.irlp.net. Each repeater, or group of amateur radio repeaters, is connected to the Internet. The technology uses streaming audio. This technology is similar to Echolink except it is RF to RF, not computer to repeater. Basically is simply connects repeaters together via the internet. An operator connects to repeater via RF the same way as regular repeater operation, with the exception of a sysop using DTMF tones to turn IRLP on or off.

Operators within the radio range of a local node are able to use DTMF tone generators to initiate a node-to-node connection with any other available node in the world. Each node has a unique 4 digit node number in the range of 1000-8999. To find the status of a repeater can be found at http://status.irlp.net. Many repeaters have Echolink and IRLP connections. IRLP runs on Linux.

W1AW (ARRL) operates its own IRLP node (4292) on 147.425 MHz (Simplex). This node is in operation primarily for EMCOMM use. However, the node is usually monitored during daily regular visitor operating times. The node is non-operational during regular broadcast times.

Portable Antenna Systems

Dr. John A. Allocca, WB2LUA
www.WB2LUA.com
3/6/16

System 1 - HF / VHF / UHF

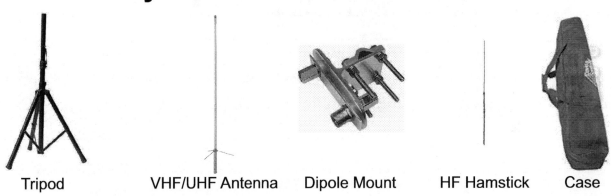

Tripod   VHF/UHF Antenna   Dipole Mount   HF Hamstick   Case

Introduction

This antenna configuration is intended to be sturdy. Guying will not be required for most applications. The height of the mast will be about 12 feet. It is easy to construct without any machining and with only a few hand tools. This system uses a combination of VHF, UHF, and HF antennas.

Tripod

The tripod used, comes in a set of two. Remove the vertical inside section from one tripod by unscrewing 3 nuts and bolts. Then, insert it into the end of the inside vertical section of the first tripod. The height of the tripod mast will be about 9 feet with a mast diameter of 1.5 inches. A 1.25 inches 3 feet aluminum tube is added because the dipole mount requires a maximum diameter of 1.25 inches. The total height of the mast is now 12 feet.

UHF/VHF Antenna

The diamond X30A antenna can be used in place of the X50A antenna. The X30A antenna is 4.5 feet tall and has a gain of 3.0/5.5 db. The X50A antenna is 5.6 feet tall and has a gain of 4.5/7.2 db. Use 1/4-20 wing nuts in place of the hex nuts supplied for faster assembly.

An alternate VHF/UHF antenna would be to use a 1 inch diameter 3 feet mast in place of the 1.25" diameter 3 feet mast and a Diamond CRM UHF mount with a mobile antenna such as the Diamond SG7900A. The Diamond SG7900A is 62 inches long and has a gain of 5.0/7.6 db and costs about the same as the X50A. Other dual band mobile antenna can also be used.

HF Antenna

Hamsticks was the original name brand for the mobile whip antenna listed below. The only antennas of this type are currently being manufactured by MFJ. A dipole hamstick configuration can be used with the MFJ MFJ-347 Dipole Mount. The bolts that come with the MFJ dipole mount are not quite long enough to fit a 1.25 inch tube. The use 50 mm long bolts are required. Fully assembled, most hamsticks extend to 8 feet long. With two hamsticks in a dipole configuration, there will be a total length of 16 feet. Hamsticks are currently available in 10 meters, 20 meters, 40 meters, and 75 meters. Hamsticks should be tuned with an antenna analyzer before normal use.

Parts List

Pair of Ignite Pro Tripod DJ PA Speaker Stands Adjustable Height Stand, $39.99

Gator Speaker Stand Bag, GPA-SPKSTDBG-50, 52 inches long, $19.95

MFJ MFJ-347 Dipole Mount, $19.95

Diamond X50A Dualband Base/Repeater Antenna, 4.5/7.2 db gain, 5.6 feet - $95

(2) MFJ-1610T, HF STICK, 10M, 3/8-24, W/WHIP, MOBILE ANTENNA - $14.95 x 2 = 29.90

(2) MFJ-1620T, HF STICK, 20M, 3/8-24, W/WHIP, MOBILE ANTENNA - $14.95 x 2 = 29.90

(2) MFJ-1640T, HF STICK, 40M, 3/8-24, W/WHIP, MOBILE ANTENNA - $14.95 x 2 = 29.90

(2) MFJ-1675T, HF STICK, 75M, 3/8-24, W/WHIP, MOBILE ANTENNA - $14.95 x 2 = 29.90

(4) 18-8 Stainless Steel Hex Head Cap Screw, M6 Thread, 1 mm Pitch, 50 mm Long

(1) Multipurpose 6061 Aluminum Tube, 1-1/4" OD, 0.065" Wall Thickness, 3 feet long

(4) 304 Stainless Steel U-Bolt, with Mounting Plate, 1/4"-20 Thread Size, 2" ID

(8) 18-8 Stainless Steel Wing Nut, 1/4"-20 Thread Size

(8) #8 flat washers

(2) CABLE XPERTS, CXP08XC50, 50 feet RG8X Coax, PL-259 Male both ends $38.95 x 2?

Above prices were seen on 3/6/16

System 2 - VHF / UHF

Light Stand   CRM Mount   VHF/UHF Antenna   Case

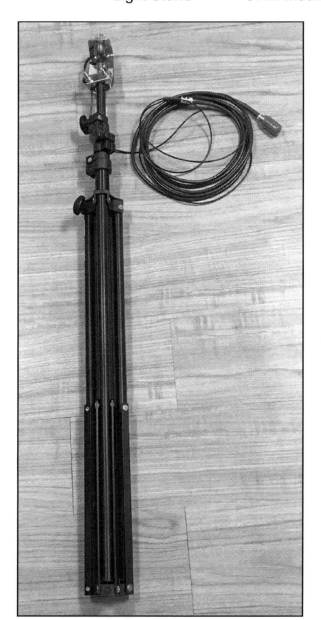

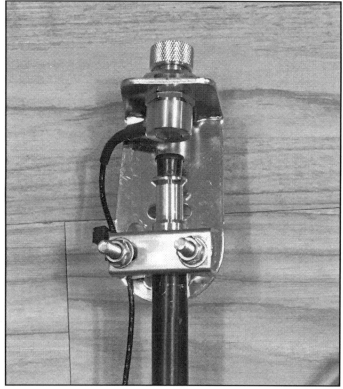

Introduction

Those who had to carry heavy radio equipment through the subways and around New York City during the Sept 11th World Trade Center incident really know the true meaning of "lightweight" and "portable." For portable use, the smallest, lightest, and the most practical of the antennas and equipment was selected. It is easy to construct without any machining and with only a few hand tools. The extra 50 feet of coax and barrel connector may or may not be needed depending upon the distance from the antenna to the transceiver.

Photographic Light Stand

The Smith-Victor 10 foot aluminum light stand extends to 10 feet high and weighs only 3 pounds. It collapses to 32" and easily fits into a 36" case. The top section has a diameter of 5/8", which is suitable for most mounting hardware. The case is long enough to accommodate the CRM mount, while permanently mounted to the top section of the light stand.

UHF/VHF Antenna

The Diamond NR7900A Dual Band Mobile Antenna was selected because of its high gain and doesn't require a substantial ground plane. NR indicates No Radial. It is 57 inches long assembled. It can be taken apart with the supplied hex wrench for travel. It has a gain of 3.7/6.4 db (2m/70cm). Other less expensive dual band mobile antenna should work as well.

Parts List

Smith-Victor, RS10, 10 ft. Aluminum Light Stand, folds to: 32 in., weight: 3.0 lbs. - $45

Neewer 36"x5"x5" Heavy Duty Photographic Tripod Carrying Case - $12.95

Diamond, C211, SO239 mount with RG8X extension cable for 16.5 feet total - $49.95

Diamond, CRM, Right Angle Bracket with U Bolts - $16.95

Diamond NR7900A Dualband Mobile Antenna, 57", gain: 3.7/6.4 db - $84.95

CABLE XPERTS, CXP08XC50, 50 feet RG8X Coax, PL-259 Male both ends - $38.95

LP, UHF-11, Double UHF Female - SO-239 - Barrel Connector - $4.99

Above prices were seen on 3/6/16

Made in the USA
Las Vegas, NV
13 October 2024

96742947R00077